SAVING SCIENCE

A CRITIQUE OF SCIENCE
AND ITS ROLE IN SALMON RECOVERY

CHARLEY DEWBERRY

First paperback edition published 2004 in the United States by
Gutenberg College Press
1883 University Street
Eugene, OR 97403

Typeset in Adobe Caslon
By Treemen Design
Printed and bound in the United States.

Library of Congress Control Number: 2004110532

ISBN 0-9746914-0-2 (paperback)

To order directly from the publishers, please send $19.95 plus add
$3.00 shipping to the price of the first copy, and $1.00 for each
additional copy. Send check or money order to:

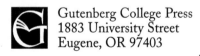 Gutenberg College Press
1883 University Street
Eugene, OR 97403

CONTENTS

PREFACE

The philosophical issues this book addresses have interested me for over twenty years, since the late 1970s when I came to Oregon State University to complete a Ph.D. in stream ecology. Eventually I transferred to the University of Oregon to pursue a Ph.D. in philosophy of science, which I completed in 1995. Both science and philosophy have been the pursuits of my life. During the last ten years I have also been a tutor at Gutenberg College, where I explore with students many of the issues at the heart of this book.

I also worked for the United States Forest Service regional lab for about nine years, and for the last ten years I have worked for the Pacific Rivers Council and Ecotrust as a salmon restoration ecologist.

I would like to thank the students and my colleagues at Gutenberg College. Much of this book is a result of our interactions over the last 10 years. I am grateful for the opportunity to continue to learn and think about science and its role in culture.

I owe much gratitude to Ron and Robby Julian who worked long and hard to transform my draft into this final form. Their efforts greatly clarified the arguments.

I thank Sarah Wierenga who drafted the text for the back cover of the book.

I thank my wife, Susie, for her support and our boys, Dylan and Andrew, for their patience while this book was being written. I hope that in some small way this book helps insure that our boys are able to fish for salmon on the Siuslaw River in the years to come.

Charley Dewberry
Florence, Oregon
June 2004

CHAPTER 1: **Introduction**

From legislators to industry people to environmentalists, almost everyone assumes that sound science is the sole basis for recovering Pacific Northwest salmonids. They agree virtually unanimously that all actions to recover depleted salmon runs should be based on it. They do not, however, agree on what sound science is and who has it. Interested parties have differing views on what science says and who has the last word. The Oregon Legislature has attempted to end this contentious debate twice. In 2001, both assemblies of the Oregon Legislature passed a bill (HB 3344) defining sound science, only to have the governor veto it. In 2003, at the request of the Oregon Cattlemen's Association, Representative Close introduced a bill (HB 2416) in which science was defined as "the systematic enterprise of gathering knowledge about the universe and organizing and condensing that knowledge into testable laws and theories." The time is long overdue, therefore, to critically examine "sound science" and its role in salmon recovery.

WHAT IS THIS BOOK ABOUT?

Most visions of sound science contain rarely articulated elements, or themes, from the two philosophies of science that dominated the early part of the twentieth century, *logical positivism* and Karl Popper's *hypothesis testing*. One dominant theme is the idea that scientific knowledge is privileged knowledge, or the only form of knowledge. Another is the idea that the scientific method results in knowledge that is more certain and objective than knowledge obtained by other means because the scientific method separates the biases of the investigator from the test; science, therefore, deals only with "objective" facts.

This book will challenge those ideas, showing how science—specifically with regard to salmon recovery—has a status exalted far beyond its bounds. But further, this book will propose that the implicitly assumed ideal view of science embraced by almost everyone cannot survive critical examination. What is commonly called "sound science" is not defensible, and scientific knowledge is neither the sole form of knowledge nor an exalted form of knowledge.

To critique a commonly held but nebulously defined concept is complex and difficult. I could approach it in a number of ways. I could articulate the commonly held positions and all their variations and then critique them one at a time. This approach has the advantage of being thorough, but it is lengthy and not very readable. Or I could identify central assumptions common to all the positions and show that one or more of these assumptions are false. Although this approach cannot deal exhaustively with all problems, it has the advantage of efficiency and readability. I have therefore taken this second approach.

I have picked three assumptions that are central to the positions of a majority of the participants in the salmon recovery debate:

1) Science ought to have a privileged position in salmon recovery.

2) Science deserves its privileged status because the scientific method gives science a certainty and objectivity denied to other pursuits, allowing it uniquely to separate facts from values.

3) The objectivity of science is further guaranteed by the formal peer-review process, which ensures the quality of the data.

Although these assumptions are widely accepted, I will show that all three are fallacious, that indeed no basis exists for accepting any of them. I will then replace these false assumptions with true ones. The outline of this book, then, is as follows: chapters one through five will diagnose the problem and critique the three assumptions; chapter six will prescribe a way to save science and establish its true role in salmon recovery.

THE VALUE OF THIS STUDY

A large number of empirical scientists and pragmatic practitioners will immediately object to my critique. In their view, nothing is to be gained by such an exercise. To them, philosophy of this sort is not useful; indeed, purging this type of philosophy from the debate has been the major advancement in the philosophy of science. For them, science has been separated from philosophy, and only an outcome or effect matters. And the only outcome they can see for the philosophical discussion I am proposing is to impede the recovery of salmon. I will acknowledge that such a discussion does result in a short-term focus on issues that seem esoteric to the matters at hand; as in restoration, however, what counts is the long-term. If I am right, at stake is not just the future of salmon, but also the future of science and its role in culture.

My short reply to these practical people is that bad philosophy is hurting salmon. At least three serious problems are hampering salmon recovery:

1) Many of the wrong people are now making critical decisions about salmon recovery.

2) Much critical information for making good decisions is kept out of the discussion.

3) The basis for decisions made by key individuals is wrong.

The immediate and long-term effects of these problems are bad for salmon and ultimately bad for science. These are serious claims, and to support them will require a longer reply—that is, the rest of this book—because the philosophical issues at the heart of this book have a profoundly practical impact on salmon recovery.

Ultimately my critique in response to the pragmatic detractors is that they themselves hold one or more of the three fallacious assumptions I listed at the beginning of the chapter. To initiate the longer reply, then, let me tell you the genesis of this book.

It began April 28, 1998, when the moderator of a meeting that was part of the Oregon Plan to recover salmon outlined the ground rules of the process. He articulated his view of sound science and while doing so demonstrated allegiance to all three fallacious assumptions I mentioned above. He implied that science was the only form of accepted knowledge; he explicitly claimed that science's unique role is to separate facts from values; and he implied that science's method gives it its unique standing (assumption two). Furthermore, he stated that the peer-review process is the ultimate ground upon which to accept a work as "science" (assumption three). Together these two assumptions provide the support for what seems to be the moderator's perspective, that a certain kind of empirical, objective science should have a privileged position with regard to salmon recovery (assumption one).

Most scientists would agree with the moderator of that meeting. None of the scientists on the podium that day objected to his directions, and no one in the audience was allowed to speak. And yet real problems beset the exalted view of science implied by the moderator. Let me illustrate them with four brief examples that show how the current view of science impacts salmon recovery. While these examples may seem isolated, they are united in demonstrating the practical implications of the prevailing view of science.

EXAMPLE I

The National Marine Fisheries Service (now part of the National Oceanographic and Atmospheric Administration) initially listed sea-run cutthroat trout as an endangered species only in part of the Umpqua River, an Oregon central-coast stream, because the only scientific data available on Oregon coast sea-run cutthroat have been collected at Winchester Dam on the Umpqua River near Roseburg, Oregon. No other scientific data had been collected which could demonstrate the need for listing. That data indicated that populations have declined and are continuing to decline.

Let us contrast this scientific information with some historical information.[1] In the early part of the twentieth century, fishermen from as far away as Europe came to catch sea-run cutthroat in another Oregon coastal river system, the Siuslaw and neighboring rivers. The Department of Fish and Wildlife considered the Siuslaw River to be the finest sea-run cutthroat fishery in Oregon as late as the 1970s, and the Oregon Department of State Lands concluded that the Siuslaw was the best sea-run cutthroat fishery in the whole Pacific Northwest. During the 1990s, however, less than five hundred wild sea-run cutthroat were caught in the Siuslaw annually.

Scientists consider these historical facts to be merely anecdotal information, however, and so the only information they use to determine historical abundances of sea-run cutthroat on the Oregon Coast is the scientific information available at the Winchester Dam, which has only been collected since 1964. By privileging scientific information over historical, they are ignoring much valuable and relevant information needed for making assessments about the status of salmon populations.

EXAMPLE 2

The view that "objective science" is the sole knowledge to be used in salmon recovery is not limited to governmental agencies. In 1995, the Pew Charitable Trust contracted a team of blue-ribbon scientists to analyze the salmon problem and provide solutions for the crisis.[2] This panel worked for two years and then submitted their report. They basically concluded that they could not determine if a salmon crisis even existed since scientific population estimates of salmon in the late nineteenth or early twentieth century were lacking, and, therefore, they did not have enough scientific data to make the determination.

Contrast the panel's finding with some historical information. At the turn of the century, Florence, Oregon, had three salmon canneries on the Siuslaw River. For about twenty years these canneries sustained an average catch of 69,000 coho salmon per year. If we assume that the catch rate was about forty percent, we can estimate that the run in the Siuslaw was around 175,000 coho. The largest runs were probably in the neighborhood of one-half million fish.

Today the number of coho salmon that return to the river—in a good year—is about 4,000 fish.[3] During the early-1990s, when there was a coho salmon fishing season, managers tried to harvest 35,000 wild coho annually from all California and Oregon streams combined. At the turn of the century, the Siuslaw River alone sustained twice that many. While this information clearly shows that coho salmon populations are severely depressed from their numbers at the beginning of the twentieth century, it is not used because it is not "scientific." Rather, it is considered second-rate, anecdotal information—useful as history but not for science.

Yet the question the scientists contracted by the Pew Charitable Trust needed answered is rightly a question for history, not science: "What were the historic numbers of salmon found in the rivers?" Science usually deals with repeatable information, such as A causes B. If the question is a historical one, however, such as what actually happened during a period of time, then the scientific method is inappropriate for the task. Historical methods aim to answer historical questions, and the methods of history are not the same as the methods of science. Scientific information may be important in determining what happened, but it plays a subservient role to historical methods in matters of history.

By limiting knowledge to scientific knowledge and limiting facts to scientific facts, a very flawed picture emerges, one that severely restricts our ability to see what happened in history. An exalted view of scientific knowledge has crippled our ability to use valuable historical information that is necessary for developing restoration plans.

EXAMPLE 3

In 1997, Governor John Kitzhaber initiated the Oregon Plan for the recovery of salmon, its goal being to devise a voluntary strategy for the recovery of Pacific salmon. If successful, the plan would make listing the salmon as threatened or endangered under the Endangered Species Act (ESA) unnecessary. A critical element of the Oregon Plan was establishing a team with final authority to evaluate the plan's success. This team, called the Independent Multiple-Disciplinary Science Team (IMST), consists—as its name implies—only of scientists.

Giving this role to a team of scientists suggests that scientific knowledge is the only relevant information for evaluating the Oregon Plan and assumes that only scientific experts can provide objective evaluation. No other group or individual has been selected to review the plan. This example illustrates how scientists have been given an exalted status with regard to the evaluation of salmon recovery, and this is so primarily because scientific knowledge is seen as more certain and objective than other forms of knowledge.

EXAMPLE 4

In the Fall 2001 *Piscatorial Press*, the newsletter of the Oregon Chapter of the American Fisheries Society, the society's president called for a scientific panel to be convened to define sound science.

But are scientists uniquely (or even best) qualified to make such a definition? I will address this question in length in chapter two; for now, however, let me just say that questions about the nature of science and its role in society are not themselves scientific questions, and the methods of science are not useful for addressing them. Traditionally these issues have fallen in the area of philosophy of science. Philosophers of science, not scientists, are best qualified to separate science from pseudo-science.

How, then, has it happened that scientists are being asked to address questions they are not best qualified to resolve? At the beginning of the twentieth century, *scientism*, a theory that assumed scientific knowledge to be the only form of knowledge, prevailed. The major proponents of this view were the logical positivists, most of whom were scientists or at least admirers of science. And positivism has, I believe, bequeathed us the idea that scientific knowledge is at least a privileged kind of knowledge. (I will discuss the legacy of the logical positivists in more detail in chapter three.)

The result of assuming that scientists have privileged information has often been that the wrong people are making decisions. Effective problem solving, including in the area of salmon recovery, involves using people whose skills and expertise are best suited to the questions at hand.

CONCLUSION

This book began in response to the ideas proposed by the moderator of the 1998 Oregon Plan meeting, ideas that I believe are counter-productive to the goals of salmon restoration. But this book is not just a critique of that moderator, nor a critique of that meeting's process. What I am proposing has implications far beyond the Pacific Northwest and salmon recovery. Salmon recovery is merely the lens through which to examine much broader issues. I selected it based on my immediate experience and my perception that the model of science and its role in salmon recovery is expanding to issues beyond resource management. At stake, ultimately, is the nature of science, its accepted methods, and its proper role in society.

1. The historical information about salmon to which I refer throughout this chapter is documented in A Watershed Assessment for the Siuslaw Basin, produced for the Siuslaw Watershed Council by Ecotrust in 2002. See <www.inforain.org/siuslaw/>.

2. Botkin, D. et al. 1995. "Status and Future of Salmon of Western Oregon and Northern California." The Center for the Study of the Environment. Santa Barbara, CA.

3. See <www.oregonstate.edu/Dept/ODFW/>, the website of the Oregon Department of Fish and Wildlife's Corvallis Research Lab, for current adult coho numbers from the Siuslaw River.

CHAPTER II: **Scientists as unskilled workers**

In 1998, when the moderator opened the Oregon Plan meeting, his remarks were very clear about science and the scientist's role in salmon recovery. The following is a partial transcript of those remarks:

> I would like to add a couple of comments about the objectives. I think that as you look at the panel here, we have a collection of scientists all with significant scientific credentials. And those scientific credentials come from their peers. These folks have a career that has been based on recognition through the peer-review process. And so it is those credentials and their wisdom that has come over their career that we would like to have them share with us today.
>
> On the other hand, I think that if you took any one of these folks out for a beer you would find that they had values and opinions. And our challenge today is to focus on the scientific credentials and to save those values and opinions for the tavern at 5:30 tonight.
>
> On the other hand, I suspect that we are going to find on this topic that the knowledge base is imperfect. And we will have to rely on the scientists to provide expert wisdom based on their scientific perspectives. So in that regard, I think that we might add one more objective and that would be to encourage an honest and forthright display of knowns, unknowns, and areas of disagreement about cumulative effects. Then as you read the remaining objectives I think that they fit pretty well. But let us make sure that we do our best to separate fact from values, opinion, or fiction about what is known about cumulative effects. Ultimately, you folks on the MOA[1] committee are going to be challenged to apply the information that is presented to you within the societal context that is full of politics, values, and opinions. You are the folks that really have the challenge to take the science and put it into the socio-political, economic context.
>
> So, speakers, I guess I'd challenge you a bit as with gunslingers in bars. Please check your values and opinions at the

door—at least initially. You will help this process the most by sticking to the facts. Tell us: what is known based on vigorous scientific tests; what is the scope of influence or extrapolability of the particular work that you will be speaking about; where is there honest disagreement about interpretations of available data. Please share that uncertainty where you can.

Okay, and remember at the end of the day, today, the MOA folks have the chance to ask questions. So be thinking about the questions that you want to ask. And that does not mean that you do not have questions earlier on, but this is your opportunity to learn.[2]

There is room for discussion about what the moderator meant, but I think it is fair to say he demonstrated his adherence to the three assumptions I listed in the previous chapter:

1) Science ought to have a privileged position in salmon recovery.

2) Science deserves its privileged status because the scientific method gives science a certainty and objectivity denied to other pursuits, allowing it uniquely to separate "facts from values, opinion, or fiction."

3) The objectivity of science is further guaranteed by the formal peer-review process, which ensures the quality of the data.

Assumption one is true because assumptions two and three are true. Although in later chapters I will show assumptions two and three to be fallacious, in this chapter I will begin my critique by assuming for the sake of argument that assumptions two and three are true. I will stipulate that the scientific method and the peer-review process provide a uniquely certain and objective kind of knowledge. The moderator at the Oregon Plan meeting assumed this, and thus he made clear that the person with "scientific credentials"—that is, the published scientist whose manuscripts have been critiqued by several of his or her peers and accepted for publication in a scientific journal—is best qualified to evaluate salmon recovery plans. Because science gains uniquely valuable knowledge by using the scientific method, science—and those who have been acknowledged for practicing such science—should have a privileged position. The logic of this claim seems clear enough on the surface. This chapter, however, will show that this claim ultimately leads to absurd conclusions. I will argue that, the so-called objectivity of science notwithstanding, science deserves no privileged position; even when we grant the validity of assumptions two and three, assumption one is *still* wrong.

THE OBJECTIVITY OF SCIENTIFIC INFORMATION

Science is currently defined as a method of investigation that, if carefully followed, yields "objective knowledge," results that are more objective and more certain than could be obtained by other means of investigation. Science's exalted or privileged status results from its unique method of inquiry, a method of investigation often termed "hypothesis testing."

The scientific method, or hypothesis testing, is a mechanical process, and it is the mechanical workings of the method that supposedly ensure the outcome. This mechanism gives science the presumed objectivity, detachment, and certainty denied to other means of knowing because the outcome does not depend on any attribute within the scientist. Indeed, the method aims to detach the scientist from the test, negating any belief or any skill that a particular practitioner has.

Given this view, science could even be done by a machine. Samples could be placed into the machine, the outcome could be a dial reading that could be compared to the expected result, and statistics could be used to decide whether the results supported or refuted a particular hypothesis. This detachment and resulting objectivity are the basis for the claim that science alone deals with "facts."

[I know I have stipulated in this chapter that assumption two is true, but I can't help noting that using statistics to determine results is an arbitrary method. There is no particular reason to set a decision break point at 90% or 95% confidence limits. The values, which are arbitrarily selected points along a continuum of values, could just as easily be 91% or 94%. Yet the knowledge resulting from the scientific method is claimed to be more objective and certain than knowledge in any other field of endeavor because the individual scientist and his or her biases have been detached from the process. Now—back to the chapter.]

The current view of science has largely come to us from the two philosophies of science dominant during the first half of the twentieth century, *logical positivism* and the philosophy of Karl Popper. From logical positivism come the ideas that scientific knowledge is the only form of knowledge and that science is objective measurement; that is, we can only have knowledge about elements that we can quantify, and our measurements must be neutral with regard to any other factor. From Karl Popper comes the view of science as hypothesis testing, and specifically, testing hypotheses in order to rule out possibilities by falsifying theories; that is, scientists, assuming that no one could ever prove a theory's truth absolutely, set up controlled experiments to try to show whether a theory is wrong. These two philosophies (which we will examine in more detail in chapter three) led to the idea that the unique value of the knowledge gained by the scientist is the objectivity and detachment by which the knowledge was gained.

Given this understanding, we can justifiably ask, Who could be better at proposing and assessing salmon restoration strategies than the scientist, the one with access to a higher kind of knowledge than anyone else? Before answering this question, however, we must ask another: On what basis can a scientist make judgments and speak about what is true? The Oregon Plan moderator claimed the scientist's qualification is based on "recognition through the peer-review process." But here we face a huge problem. Activities such as making judgments, diagnosing, evaluating, and proposing solutions all require skill, and yet the very method that ensures the unique objectivity and detachment of scientific knowledge—the mechanical nature of the process—requires that the scientist's skill be irrelevant to the results.

Just as a cook would follow a recipe in a cookbook, scientists mechanically follow the scientific method, which sets aside all experience as bias to be removed from the process. This is the basis of science's greater claim to objectivity. Reproducible results are a fundamental characteristic of science; scientists assume that *any* practitioner who adheres to the same method will get the same results, just as any cook who follows a recipe should get the same results. It is the mechanical working of the method—not any attribute of the scientist—that is important.

If adherence to the scientific method is what gives scientific knowledge its greater privilege, then the scientist has no superior basis for making any scientific judgments. Scientists become, in essence, unskilled workers. True, they are knowledgeable workers; they are disciplined workers. But the supposed virtue of the method, the superiority of scientific knowledge, arises from the fact that the results do not depend on the individual, biased attributes that the exercise of skill demands. Making judgments is a skill, but the supposed superiority of scientific knowledge depends on skill being irrelevant. Thus the dilemma: how can an unskilled worker use skills? To help unpack this dilemma we need to take a careful look at the nature of skills.

THE NATURE OF SKILLS

Because the scientific method is designed to work against our everyday experience, one might argue that following it carefully requires a great deal of skill. While I agree that doing good science within the framework of the scientific method takes careful work, I would not call it skill. To make such a claim is to misunderstand the nature of skills. Following the scientific method, like following a recipe in a cookbook, is very different from performing skillfully. Following a recipe involves paying close attention to each step in linear order. Performing a skill involves much more.

A person learning any new skill—whether it be shooting baskets, riding a bicycle, playing a musical instrument, identifying an organism, or diagnosing a patient's health—faces a thousand "rules" that he or she must follow simultaneously. For example, when a boy learns to ride a bicycle, his parent tells him everything he needs to do. Initially the boy focuses on one or more of the rules, but his attempts to ride are awkward because he cannot follow all the rules at once. As he practices (that is, as he focuses on doing particular rules), he gets more proficient. In time, following the rules becomes second nature to him, and he learns to follow a number of unarticulated rules as well. Eventually the boy no longer focuses on the rules; he just rides the bike. At this point, he has the minimum ability to perform the skill of riding a bicycle. In some sense, he is following all the rules, but he is not focusing on any one of them; he is just riding the bicycle. If the boy focuses on the rules again, then he has ceased doing the skill and is only practicing. Performing a skill, then, depends on the practitioner; it is an act of both "knowing" and "doing."

A skill is an achievement, which a person accomplishes by observing a set of rules he or she did not previously know.[3] Now, the boy riding the bike does not know *all* the rules necessary to ride. He cannot articulate how he rides. He does not know how he balances. He does know that if he is falling to the right, then he must turn the handlebars to the right; but he does not know that turning the handlebars results in a centrifugal force that pushes him back to the left and keeps him from falling. Even though the boy does not know the explicit "rule" behind the maneuver, he can execute it without conscious thought because he has achieved the skill of riding.

SCIENTIFIC JUDGMENT AND SKILL

As we have seen, skill depends on a practitioner's personal knowledge of how to do something. In contrast, the goal of the scientific method is to eliminate personal knowledge—that is, to detach human biases from any results—in order to achieve the objectivity that supposedly makes empirical science unique and gives it greater standing. Thus, those holding the current view of science would label slavishly following the prescribed method (much as a cook would follow a recipe) "science" and performing skillfully "art."

There can be no art in science—or so everyone seems to be saying. In order to gain objective knowledge, science must be a strictly mechanical process that precludes any skill. By thus elevating objective knowledge, we remove personal knowledge (including skills) from our accepted body of knowledge. Only the objective, peer-reviewed work remains.

If science is the mechanical process I have described, then no judgment exists within science. Only choices that can be determined mechanically by the scientific method are acceptable. Furthermore, any reference to scientific judgment is without content because making a judgment is a skill; and skill, belonging as it does to the realm of personal knowledge, is an art. When a person makes a judgment, he or she cannot construct a complete decision-matrix to determine the results mechanically because he or she is not conscious of all the factors that go into making the judgment nor how those factors are weighed. If we say the scientific method yields the only acceptable kind of knowledge, then we are saying knowledge has nothing to do with judgment. In fact, we are ultimately saying that knowledge has nothing to do with intellectual reasoning. The scientific method is based on empirical tests using statistics, and the purpose of using statistics is to remove any intellectual reasoning, including judgment, from the process.

We can now see the difficulty posed by the moderator's claim at the Oregon Plan meeting that a body of peer-reviewed work is the basis upon which the researchers could speak. Presumably, work based on empirical tests using statistics that has been reviewed by peers best qualifies a person to make judgments about salmon recovery strategies. Why? Judgment and intellectual reasoning have nothing to do with examining hypotheses by statistical methods; statistical tests remove judgment and intellectual reasoning from consideration. What basis remains from which a scientist can comment?[4]

We have come to the absurd conclusion, then, that making scientific judgments, which is an exercise of skill, are best made by those whose work excludes skill as a criterion. We can clearly see the absurdity of this position by making an analogy to medicine. When we are sick, do we go to the medical researcher or to the general practitioner for a diagnosis? The medical researcher is a scientist who tests hypotheses and does statistical analysis on the results and who, therefore, supposedly has superior, objective, unbiased knowledge gained through a mechanical process. And yet it is not the medical researcher but the general practitioner to whom we take our aches and pains.

Diagnosis is making a judgment, a skill. The basis upon which the general practitioner makes a judgment is the years of experience he or she has had making judgments—that is, his or her personal knowledge. Although the researcher may know all the literature on a particular topic—that is, the researcher has objective knowledge—he or she may not have made a single diagnosis since entering the research-side of medicine, and, therefore, the researcher likely does not possess the general practitioner's skill of general diagnosis. In fact, doing science, which detaches skills from the process, works against the development of skills. Over time, the general practitioner's diagnostic skill increases, but the researcher's diagnostic skill erodes, which is why we go to a general practitioner rather than a medical researcher when we are sick.

Analyzing a patient is very different from analyzing the statistical outcomes of various tests. The general practitioner and the medical researcher might rely on the same tests, but they evaluate them differently. We have not tried to replace the general practitioner with a written mechanical decision-matrix, nor do we ask for the one the practitioner used to determine our diagnosis. In fact, he or she could not construct a complete decision-matrix because the practice of diagnosing the health of a patient can never be reduced to a mechanical process.

To claim that only the medical researcher has any knowledge about medicine or that only the researcher is qualified to make judgments in medicine is the death of medicine. Such a claim denies the role of the general practitioner and denies that personal knowledge is true knowledge. The method of the general practitioner is not scientific; the practitioner's work is based on his or her skill, and skill is outside the realm of science. Yet, if knowledge in medicine is limited to the objective knowledge of the medical researcher, then medicine can no longer diagnose because diagnosis is a skill and not objective knowledge; and furthermore, judgment cannot exist within medicine because making judgments is a skill based on experience and therefore lies outside of science.

With our medical analogy in mind, let us look at salmon restoration. Who is qualified to diagnose and prescribe in this situation? According to the prevailing view, only the scientist is allowed to provide the facts and to make judgments about the recovery of salmon. The superior, objective, unbiased knowledge is the knowledge found in the peer-reviewed science literature. Therein lies the problem. Diagnosis is a skill. But to the extent that scientists have faithfully practiced the scientific method, they have spent their lives setting aside skills, including judgments. They are, in fact, handicapped when it comes to making judgments. On the other hand, people who have spent their entire lives working and living with salmon—those who have developed skills that could be the basis for judgment—are viewed as having only personal knowledge about salmon. According to the prevailing view, those who have had the least opportunity to develop judgment skills are seen as more qualified than those who have had the most opportunity to develop those skills. The moderator at the 1998 Oregon Plan meeting portrayed both establishing the facts and making scientific judgments as the job of the peer-reviewed scientist, whereas values and skill would enter into the process only in the wider debate about how to set policy. And, to my knowledge, no one has criticized the moderator's assumption until now.

SCIENTIFIC JUDGMENT AND EXPERIENCE

As we saw in the example of the medical doctor, making judgments is a skill, and skill is developed through experience. Experience, then, is a foundational requirement for making good diagnoses. As we have also seen, however, the scientific method tends to eliminate skill as a factor; thus having a body of peer-reviewed work may not indicate that a researcher has any experience making judgments in the field.

In fact, the problem is potentially more serious than I have indicated. Since the scientific method is designed to detach the scientist from the process, how better to do that than to have other people collect the data? Consider the following scenario; it is not unusual. For a major research project, the lead scientist writes a grant proposal in which he or she outlines a hypothesis or what is to be done, the methods that will be used to collect the information, and the statistical methods that will be used to analyze it. When the scientist gets the grant, he or she hires technicians to collect the data. The scientist then prepares data forms and spends time in the field—often a day or less—showing the technicians how to collect the data and enter the information on the data form. This may very well be the only time the scientist sees the project in the field. The "real" scientific work is setting up the test properly, which the scientist has done, and then analyzing the data statistically and writing the report. The scientist, however, may not even do the statistical analysis; he or she may only write the final report. And if the scientist has good technicians, the technicians may even draft the manuscript.

Since the method of science—not the scientist—is important, whether a scientist actually does any of the work on a project is irrelevant. When all the lead scientist does, then, is write the grant proposal and get the money, on what basis does he or she speak for the results or make judgments about it? Indeed, the scientist has very little basis on which to speak about the project because the basis on which someone can make judgments is experience, and the primary researcher often has little experience with the data collection.

Understanding this implication of the objectivity of the scientific method has helped me understand two of the more troubling experiences of my graduate education in fisheries and wildlife at Oregon State University. The first experience was seeing the emphasis on statistics over experience. On the one hand, I had always assumed that because a scientist's judgment is built on years of field experience, field experience was the most important element of good science. I therefore emphasized field experience as part of my education; I never missed an opportunity to spend time in the field, and upon hearing the results of a recent experiment, I would sometimes respond that it did not make sense given my experience. I would be told, however, that my experience was only anecdotal; it was not factual or scientific. On the other hand, my graduate committee wanted me to take more than half of my doctoral program in statistics. I fought it, but still I took more than two years of it. My point is not that statistics do not have a role to play in science, but that the role is minor compared with the experience and skill gained in the field. When I was in graduate school, a student could complete the course work in stream ecology without taking a single field course. (Several courses included a field trip, but that hardly constitutes field experience.) To propose taking field courses and no statistics, however, would not have passed the laugh test. Completion of a thesis project was the only field experience deemed necessary for a scientist; experience and personal knowledge were not considered necessary for scientific knowledge.

The second experience that troubled me in graduate school was seeing how little actual field experience, or personal knowledge, many researchers in salmon-related fields had and then discovering that this lack of experience was not seen as a problem. Today some of these scientists with little field experience are on scientific panels making critical evaluations and decisions about salmon recovery. Some nationally and internationally known scientists who publish literature on salmon have not conducted extensive fieldwork since they completed their degrees or post-graduate work, and some cannot even identify the fish in the field. This is not the case for all scientists on the panels, but a number of them have little field experience with salmon. In fact, the ability to identify juvenile and adult salmon in the field is not a prerequisite for membership on these committees. My point is that no one sees this as a problem.

For example, consider Dr. Daniel Botkin, who chaired the blue-ribbon science panel that investigated the Pacific Northwest salmon issue for the Pew Charitable Trust mentioned earlier. Dr. Botkin does not claim to be a salmon biologist; he is an ecologist. At the same time that he chaired the Pew committee, he wrote a natural history book in which he mentioned seeing twenty-pound coho salmon spawning in an Oregon stream.[5] I have worked in Oregon for over twenty years, and I have never seen a coho that weighed over fifteen pounds. Although coho salmon can get that big, it is doubtful that one grew that large in a year with very poor ocean conditions, and Dr. Botkin did not indicate that these coho were especially large. Chinook salmon, on the other hand, weigh twenty-two pounds on average. I hope Dr. Botkin's reference to the spawning "coho" salmon was a typographical error, rather than his having misidentified the fish or overestimated its weight by up to fifty percent.

I refer to Dr. Botkin not to pick on him, but rather to highlight this fact: as things stand, there is no reason that he should know anything about salmon biology other than the "objective" knowledge in the scientific literature. He was hired as a scientist, and it does not matter if he knows anything about salmon in the field. The only relevant question is whether he can evaluate the scientific information and literature, which are considered the sole source of knowledge we have about salmon and which, therefore, provide the only legitimate facts to be considered—if, of course, we assume the prevailing view of science.

JUDGMENT AND PERSONAL KNOWLEDGE

Thus far we have seen that assuming science ought to have a privileged position in salmon recovery leads to absurd conclusions: the diagnoses of those whose work de-emphasizes skill are to be preferred to those who have skill, and the diagnoses of those whose work de-emphasizes experience are to be preferred to those who have experience. Now, let us return to our examination of skills for a moment and the example of the boy riding a bicycle to discover the final absurdity of assuming an exalted view of science: the bizarre view of knowledge that must result.

Physicists and engineers have determined the specific physical laws (the objective knowledge) necessary for riding a bicycle. But what if a particular physicist who can articulate these laws cannot ride a bicycle? Does this physicist have a greater understanding of bicycle riding than the boy who, using personal knowledge, just rides the bike?[6] While I would acknowledge that the scientist knows something that the boy does not know explicitly, I would deny that the scientist's "objective" knowledge of physical laws is a higher or more desirable kind of knowledge. And, therefore, I would also deny the logical positivists' claim (which pervades the prevailing view of science) that objective knowledge is the only form of knowledge.

To claim that objective knowledge of the laws of bicycle riding is the *only* knowledge of bicycle riding is wrong. It is absurd to claim that our non-riding physicist has knowledge of bicycling riding but that our boy bicyclist has no knowledge of bicycling riding just because he cannot articulate certain physical laws. The notion of objective, more certain knowledge is largely irrelevant in this case; the relevant knowledge is the skill of bicycle riding, or personal knowledge. The physicist has a very restricted and not very useful knowledge of bicycle riding; he or she has no skill. The bicyclist may not be consciously aware of the physical laws governing bicycle riding (although he must "know" them in some sense in order to ride), but he can use them; he has skill.

Likewise, to accept only the research scientist's objective knowledge when considering issues in salmon restoration is wrong. Doing so relegates the field experience (the personal knowledge) of workers and fishermen who have spent major portions of their lives working with salmon on a daily basis to opinion, to second-rate anecdotal information. Such an exalted view of empirical science and its objective knowledge cripples our restoration efforts because the research scientists often lack basic experience with salmon and the resulting skills that are crucial for making intelligent judgments. In fact, as we have seen, the scientific method itself mitigates against the development of that experience and skill.

According to the view of science accepted by almost everyone, experience and time in the field are irrelevant; these are biases best eliminated from the scientific process. An implication of this view would seem to be that the best scientists are those with the least experience—they have fewer biases to overcome. I do not agree.

CONCLUSION

Who can best provide facts and make judgments about salmon restoration? The moderator quoted at the beginning of this chapter believes this job belongs to scientists whose work has been reviewed by their peers. Why? It is because peer review ensures that these scientists adhere to a method that yields objective knowledge. The requirements of the scientific method, however, defy the very nature of making judgments. Diagnosing problems and recommending solutions are skills that require personal knowledge and experience. Yet, the supposed superiority of the scientist's knowledge is grounded in the objectivity gained by a method that aims to eliminate personal knowledge and experience and, thus, skill. An absurd situation results. By elevating the "objective" knowledge of the scientist we diminish the importance of personal knowledge, the very knowledge most critical to the skill of diagnosing and solving the salmon problem. Therefore, the prevailing view that science and objective scientific knowledge ought to have a privileged position in salmon recovery cannot be true.

1. A memorandum of agreement (MOA) signed by former Oregon Governor Kitzhaber and National Marine Fisheries Service (NMFS) Regional Administrator Will Stelle established a number of terms and conditions for collaboration between Oregon and NMFS in the implementation of the Oregon Plan to recover salmon species listed under the Endangered Species Act.

2. I transcribed the moderator's comments from a tape of the meeting provided by the MOA committee.

3. This analysis of skills is based on Michael Polanyi's work in *Personal Knowledge* (Chicago: University of Chicago Press, 1957), 49.

4. At the beginning of the development of science, Francis Bacon, the father of modern empirical science, made a different but related argument for scientists being unskilled workers. As Bacon saw it, relegating reason and learning to a position subordinate to the senses significantly lowered the intellectual qualifications required for scientific work. In fact, Bacon argued that anyone who had hands and eyes could perform experiments. The work could be accomplished by a large number of equally unskilled people; it would be democracy in action. In Bacon's view of empirical science, this is how scientists would conduct their work.

5. Botkin, D.B., *Our Natural History* (New York: Putnam and Sons, 1995), 178.

6. Michael Polanyi raised this insightful question in his discussion of the role of skill in science. (*Personal Knowledge*), 49.

CHAPTER III: A critique of objective science

"Well I suppose that the most important of the defects was that nearly all of it was false."

—A.J. Ayer commenting on the failings of logical positivism[1]

In the previous chapter, we started our analysis of the three assumptions discussed in chapter one:

1) Science ought to have a privileged position in salmon recovery.

2) Science deserves its privileged status because the scientific method gives science a certainty and objectivity denied to other pursuits, allowing it uniquely to separate facts from values.

3) The objectivity of science is further guaranteed by the formal peer-review process, which ensures the quality of the data.

As our starting point, we granted for the sake of argument assumptions two and three; that is, we assumed that the scientific method yields uniquely certain and objective knowledge, and we allowed that this objectivity is further ensured by means of the peer-review process. Even if these assumptions were true, however, we discovered that assumption one, granting science a privileged standing, ultimately leads to absurd conclusions. Diagnosing and solving problems is a skill, and yet if the "objective" knowledge of science is made the *sole* knowledge, no skills can be brought to bear on issues like salmon recovery.

Now we must turn our attention to assumptions two and three themselves, which, I will argue, are not true at all. In particular, this chapter will examine assumption two. Although the idea that the scientific method yields uniquely objective knowledge is very widely held, this chapter will argue that this idea is wrong.

Many scientists, including the moderator at the 1998 Oregon Plan meeting, believe the scientist's role is to provide facts. The moderator believed that science would establish the facts about salmon that the members of a larger committee could then use to make recommendations. Only scientists were allowed to input information into the political process because, according to the prevailing view, only scientific knowledge yields objective knowledge—that is, knowledge viewed either as the *only* form of knowledge or as *qualitatively different* from other forms of knowledge. Either way, scientists were (and are) considered uniquely suited to deal with the facts because they alone use the scientific method, a method that separates facts from values and thus allows scientists to be detached and objective in ways that others are not. Given this view, it follows logically that only people with certain scientific credentials were permitted to provide information for the committee to use in their deliberations about salmon recovery.

How did this widespread view of the unique objectivity of scientific knowledge come about? I am aware of two versions of the argument for the certainty and objectivity of scientific knowledge—what we might call the "strong" and "weak" forms of the argument. The logical positivists made the strong form of the argument, and I call it "strong" because the positivists argued for the total objectivity of scientific knowledge; they viewed scientific knowledge as uniquely certain because it was built totally on objective facts and not at all on subjective values. The weak form of the argument does not go so far—it acknowledges that values do enter into the scientific process—and yet it still grants scientific knowledge a unique status because the *testing* of hypotheses is value-free. This form of the argument was made by Karl Popper. I will critique both perspectives in turn, first that of the logical positivists and then that of Karl Popper.

28

CRITIQUE OF THE LOGICAL POSITIVISTS

According to the strong form of the argument for the unique objectivity of scientific knowledge, science has a unique status because scientific knowledge alone is objective—that is, value-free—and the results generated by slavishly adhering to the scientific method are objective facts. This view of scientific knowledge was the cornerstone of most versions of logical positivism, the most fashionable philosophy of science throughout most of the twentieth century. The positivists believed science alone dealt with facts while all other endeavors to acquire knowledge were fraught with subjective values. Therefore, many positivists believed that scientific knowledge was the only form of knowledge; all other forms of "knowledge" were mere opinion.

In the following section, I will present three criticisms of the view of science advocated by the logical positivists. The first two critiques are relatively straightforward and simple. The third critique is more involved, and I must develop it in some detail.

FIRST CRITIQUE: COMMON PRACTICE

According to the logical positivists, science is "value-free," but in actual practice it is not. For example, when I am asked to review manuscripts submitted for publication, one of the criteria for evaluating them is always the question, "Is a paper interesting or significant?" This criterion is a value, not a fact. Using a value judgment to screen submitted papers thus smuggles values into the scientific literature. Furthermore, scientists themselves pick their hypotheses based on values. Without values there would be no criteria for selecting a particular hypothesis; scientists would end up randomly investigating the most common phenomena on the planet, only rarely landing on an interesting, valuable hypothesis.

A truly value-free science implies that no hypothesis is

inherently better than any other; no criteria exist to rank or prioritize hypotheses, and therefore all hypotheses are equally valuable. A truly value-free science, then, is analogous to a camera that sees everything and nothing. Only the photographer who values some things over others can communicate something in a photograph. So, for example, in a truly value-free science, investigating whether streams whose names begin with the letter "S" are more productive for salmon would be as worthwhile as any other hypothesis, and the only criterion someone could use to evaluate any manuscript would be whether or not the reported results were achieved by mechanically following the scientific method. In the real world, however, even if the testing of "The-Letter-S" hypothesis adhered strictly to the scientific method, reviewers would rightly reject the manuscript reporting the results on the basis of values, namely, common sense and experience. Common sense and experience are not facts in the sense required by a positivist view of science. Indeed, the purpose of statistical procedures in science is, in part, to remove these very human elements; and yet both common sense and experience are smuggled in from outside the articulated scientific method. Thankfully, when it comes to selecting hypotheses, no scientists really attempt to follow the demands of a truly value-free science (even though they may pay lip service to it) because it would only lead to a morass of isolated facts and tests of mostly irrelevant hypotheses.

The science *practiced* by individuals (those working on salmon recovery, for example) does not actually separate facts from values; science as practiced does not rely only on facts. Furthermore, no scientist could follow the articulated method in a way that would lead to objective knowledge, and the scientific literature itself is clearly not value free.

SECOND CRITIQUE: **MAKING CONNECTIONS**

The logical positivists never specified the rules of induction—that is, how facts are linked together. This "connection" problem has existed in modern empirical science since its conception by Francis Bacon and others. It is one thing to come up with the facts; it is another to link them into some kind of coherent picture or theory. The positivists saw everything as either a fact or a value. Given this view, therefore, the linkages must be values because they are not facts; they are not just observed. Furthermore, since there are no established rules for the process of linking facts into a coherent picture, it cannot be objective and detached and totally reliant on facts.

One might think to get around this problem by arguing that there is no picture or theory that links and organizes the facts; there are only isolated facts, and science is simply what accumulates them. Many who hold a positivist view of science favor this position that reduces a theory's status to nothing more than a summary of the facts. This argument, however, does not describe the real world. In the real world, human beings—whether we want to or not and whether we admit it or not—link facts together to form a coherent, whole picture. To do so, we always rely on our experience and prior beliefs, and this effort comes completely from outside the scientific method. In practice, then, objective scientific facts are always linked together in theories, and since those theories are not "facts," they must be values.

THIRD CRITIQUE: **HUME'S EPISTEMOLOGY**

According to the logical positivists, scientific knowledge is uniquely certain because it is built totally on objective facts. But there are no such things as objective facts. Because this criticism of the positivists' view of science is more technical, I must first provide some background.

To the positivists and other tough-minded empirical scientists, observation is just opening one's eyes and looking. Facts are simply the things that happen—hard, sheer, plain, and objective:[2] someone sets up an experiment and gets a number. The positivists derived their epistemology explicitly from philosopher David Hume; they agreed with Hume that matters transcending human experience have no meaning and even if they had, they could not be shown to be true. Hume's empiricism can be summed up in two propositions:[3]

1) All our ideas are derived from impressions of sense or feeling. That is, we cannot conceive of things different in kind from everything in our experience.

2) A matter of fact can never be proved by reasoning *a priori*. It must be discovered in, or inferred from, experience.

According to this view, facts are the simple impressions of sense that arise in the mind from unknown sources. Once we have perceived the simple facts, then we can arrange and rearrange them, but initially we experience them passively; the mind is not active during this stage; the impressions just appear. For example, when I experience the color red and a particular shape, I call the combination of these two simple impressions a "strawberry." To claim something is a "fact," then, is to claim it can be traced back to impressions. A fact's empirical pedigree therefore determines whether or not we are justified in believing it; if there is no impression, then there is no fact.

A number of important implications arise from accepting Hume's epistemology. We only know the ideas in our mind because they are all we can experience. Therefore cause-and-effect refers to ideas in our mind, not something external to us. For example, we think the sun will rise tomorrow because that thought is a habit of our mind; there are no physical laws or other explanations for the sunrise, and the sun is just as likely not to rise tomorrow. What we mean by causation is that two ideas appear one after the other in our minds (what Hume calls "constant conjunction" or "correlation"), and that is all we know; we can never speak of causes as if they were in a world outside of our minds. We do not even know if a world exists outside of our minds or what it might be like. Likewise, other people and our own bodies are just ideas in our minds; we have no other knowledge of them or us. We do not know if rivers exist, or salmon, or anything else outside of our minds. (I will address this skepticism in the next chapter.)

The positivists were aware of these implications of Hume's epistemology, and they accepted them. I see four problems, however, with the positivists' view of facts as objective and empirically justified. None of the following observations are uniquely mine; I suspect they have been made many times in many different circumstances, and I will acknowledge my debt if and when I remember who first called them to my attention.

PROBLEM ONE:
An erroneous theory of observation/perception

The first problem with the idea of facts as objective and empirically justified is the faulty view on which it rests: Hume's (and others') erroneous theory of observation/perception. If Hume's theory of observation/perception falls, so does the claim for "objective" knowledge. The basis for the radical empiricist's claim is that impressions are facts. Impressions just happen (they either occur or do not), and because nothing mediates them, they are always reliable. If, therefore, two researchers run the same experiment, they should get the same dial readings; the dial readings either are the same or they are not, and nothing in the researchers can affect the outcome. Since the 1950s, however, philosophers of science have found Hume's observational/perceptual the-

33

ory to be in error, and two examples they cite that are most damaging to Hume's theory are the cases of ambiguous figures and inverted lenses.

In his book, *Patterns of Discovery*, N.R. Hanson[4] presented a number of ambiguous figures, which some people see as one thing (an antelope, for example) and others see as something else (a rabbit, for example). Some people can only see one or the other figure. Some can move from one figure to the other. In all cases what strikes the retina is the same, yet the objects the observers see are clearly different. The observers are not aware of seeing a series of lines, which they would all see the same; they are only aware of seeing a figure, and those figures differ. Hume's observational/perceptual theory, however, is based on the idea of bare "observation"; that is, the series of lines is perceived, and then the mind interprets it. For Hume, perception/observation is a mechanical process: first the impression occurs, and then the whole or parts of the impression can be copied, moved, and stored within the mind. The observers, however, do not report seeing the lines; they only report seeing figures. One could argue that the lines are seen subconsciously and the observer moves quickly to interpretation; however, this does not solve the problem. From an empirical point of view, what is the difference between two observers not being aware of a base perception and there being no base perception at all? Empirically, these two cases cannot be separated. These examples of ambiguous figures call into question the assumption that observation is a simple, mechanical process not mediated by past experience or other beliefs. Said another way, they call into question the idea that facts are just observed, that they just "happen" to the observer.

The case of the inverted lenses is even more damaging to Hume's observational/perceptual theory. A pair of glasses with lenses that invert an image on the retinas is placed in front of an observer. At first the observer is completely disoriented; with a little time, however, the observer learns to see normally. In fact, with a little time, the observer is not consciously aware that the image is inverted on his or her retinas; the observer sees the image just as someone who is not wearing the glasses would see it. At this point, the sense data are different for the observer wearing glasses and the one not wearing glasses, but the two observers see the same thing. If we passively perceive the sensations first, as Hume's theory states, why does the observer wearing the glasses not see the inverted image and then have to interpret it actively each time?

Philosophers of science have rightly rejected the Humean theory of observation/perception for about fifty years. The radical empiricists' claim that facts are just passively observed is false. Rather, the mind is active during observation/perception, and, therefore, what we know (all our accepted beliefs) affects what we see. The radical empiricist who wants to claim that facts are just observed objectively is wrong. There are no objective facts in this sense. While the radical empiricists see science built one objective fact at a time, their critics see each fact as "theory-laden," that is, already imbedded in a theory and experience.[5] Furthermore, implicit in this criticism of Hume's theory is a rejection of the radical empiricist's claim that the nature of observation is mechanical. Rather, observation/perception is a skill that the observer has honed throughout his or her life. It is not a mechanical process; it is high art that can never be made objective. What individuals see depends on their experience.[6]

PROBLEM TWO:
An erroneous view of theory change

The second problem with the idea of "objective" facts is closely related to the first, and philosopher Thomas Kuhn proposed it in the 1950s.[7] While the philosophers discussed above saw facts imbedded in theories, Kuhn saw them embedded in "paradigms." Facts, according to Kuhn, are only understandable within the context of a paradigm.

From a positivist view, science progresses as isolated, value-neutral facts are accumulated. Once established, facts are the "unit of currency" for science; they do not change. And because progress in science is seen primarily as the accumulation of facts, a theory changes only when a new theory explains all the existing facts better than its predecessor. Science, therefore, progresses linearly, which is why science can uniquely separate fact from values: it is based only on the facts, pure and simple.

Kuhn disagrees with the positivists. For him, theory change is not based primarily on the accumulated facts, and it is not the logical, rational, and mechanical process that positivists believe it to be; rather, it is a paradigm shift. Science is not a lone scientist working in isolation, but a community of workers with a set of shared commitments; and as a science matures, it develops a "paradigm"—that is, a set of questions, methods, and case examples that guide the community of workers in their work. But a number of cases are hard to bring into the existing paradigm, and thus ordinary day-to-day science is problem solving; the scientist's work is to figure out how to bring these difficult cases into the paradigm. The more difficult and protracted the problem, the greater the prestige the scientist who finally solves it gains from his or her peers. Over time, however, the number of difficult cases that do not fit into the paradigm increases, and a "crisis" arises. At this point, a new (usually young) worker from outside the current paradigm proposes a new paradigm, which has a new set of questions, new methods, and new case examples. Individuals embracing competing paradigms cannot even communicate with each other because they are "speaking a different language" (a different language game); they have "a different way of life," to borrow a phrase from Wittgenstein. According to Kuhn, therefore, theory change in science is like change in other arenas of life: it can be revolutionary.

In Kuhn's model, a new paradigm does not have to incorporate all the facts of the previous paradigm. Indeed, facts that were seen as critical and fundamental in the old paradigm may not even have standing in the new one. The accumulation of facts, therefore, is no longer the benchmark of science.

According to Kuhn, there are no criteria to decide logically and mechanically if the new paradigm should be embraced; there is no objective standard by which to judge competing paradigms. Scientists must base their decision to accept the new paradigm on other grounds. Somehow, though, a critical mass of workers accepts the new paradigm, the new paradigm replaces the old paradigm, and a new community of workers becomes the body of scientists.

In Kuhn's view, science has no unique status. It is like other endeavors of life, with no more ability to separate fact from opinion than any other endeavor. According to Kuhn, the claim made by the positivists and others to have objective knowledge is false. In this much I agree with Kuhn; in the next chapter, however, I will discuss how I disagree with him.[8]

PROBLEM THREE:
An erroneous view of knowledge

The third problem with the idea of "objective facts" is the Humean view of knowledge it assumes—that is, human beings start with a "blank slate," and all their knowledge originates as impressions. If we start with a blank slate, however, how and on what basis could we respond to the first impression? We cannot know anything about it; there is no reason to think the impression is significant, no way to understand its meaning, and no reason to accept it. Given this blank-slate theory of knowledge, the radical empirical project could never get started.

To help make this criticism clear, think of the mind as analogous to a computer.[9] If someone starts up a computer that has a blank slate—no Read-Only-Memory (ROM)—it will never work. The computer will not recognize the inputs from the keyboard nor have any idea how to understand the keystrokes. ROM tells the computer that inputs are significant and how to understand them, and without those initial instructions the computer is useless. Similarly, the human mind must have "instructions" for understanding its first impression; it is not the blank slate Hume posits.

PROBLEM FOUR:
A philosophy unworkable in practice

The last problem with the concept of "objective" facts also relates to Hume's radical empirical philosophy that says we cannot know anything outside of our minds. Although I cannot prove it, I do not believe that any scientists involved with salmon recovery actually believe and accept the implications of Hume's theory. On the contrary, they believe that real streams and real salmon exist in the world and not just in their minds. They also believe that cause and effect is an attribute of the external world and not an attribute of the contents of their minds. And they believe that the external world and other people exist, as do their own bodies. Not for a minute do scientists live or practice their science holding the skeptical beliefs demanded by Hume's system.

The seemingly universal practice of scientists, therefore, denies the philosophical foundations for the positivists' view of science. The idea that scientific knowledge alone is certain because it deals with objective facts (other kinds of knowledge being less reliable because they deal with values) comes to us from Hume by way of the logical positivists. Yet the practice of scientists themselves implicitly rejects Hume's epistemology, the philosophical basis for the idea that science deals uniquely with value-free and objective facts.

CONCLUSION TO THE CRITIQUE
OF THE LOGICAL POSITIVISTS

We have seen that the positivists' view of science—that is, the strong form of the argument for the objectivity of scientific knowledge—is in error. Practicing scientists in fact often interject subjective values into their science; subjective values must enter into the process of linking facts into coherent theories; and Hume's epistemology, embraced by the logical positivists in developing their view of science, is theoretically and practically unworkable.

Critique of Karl Popper's view

Perhaps most practicing scientists would not recognize their own position in the strong form of the argument for the objectivity of scientific knowledge. They would agree that values enter the scientific process when a hypothesis is selected and the manuscript is peer reviewed. Yet they would still argue for a unique objectivity to the scientific method. And to do so, they would use the weaker form of the argument for the objectivity of scientific knowledge: namely, that the aim of the scientific method is to eliminate values from the *test* of the theory or hypothesis.

According to the major proponent of this view, Karl Popper, science is testing hypotheses, and the scientist's goal is to attempt to falsify theories by proposing bold hypotheses. In Popper's view, a theory can never be proven true; it can only be falsified. If a scientist finds even one case in which a theory is false, then the theory is not true. Currently accepted theories, then, have been tested but not yet rejected.

Popper's system contains a far more sophisticated view of the interaction of theory and fact than we found in the positivists' system. For the positivists, a theory is just a summary of the facts; it can be nothing more. For Popper, two theories can present two different views and both account for all the facts. The scientist's goal is to design a crucial experiment that can test between the two theories, and the scientist's motivation is to try to reject the currently held theory.

In the following section, as I did for the positivists, I will present three criticisms of the view of science advocated by Karl Popper.

FIRST CRITIQUE: *Who is a scientist?*

Popper's view defines science in a way that would exclude many individuals I consider to be great scientists: Copernicus, Galileo, Kepler, and Newton, for example. None of these individuals primarily tested hypotheses; they were primarily restructuring our theories of the heavens. To define science in a way that excludes these individuals cannot be right.

SECOND CRITIQUE: *The problem with falsification*

Popper's method based on *falsification* of hypotheses or theories fails for the same reason that he claims that *verification* of hypotheses can never be certain. In his view, a theory can never be *verified*; any number of positive responses can never prove with certainty that a theory is true, because the possibility always exists that we have not examined a crucial case. But, in his view, *one* negative case proves that a theory is not true and that we should reject it; a theory can only be *falsified*.

Consider, however, the following case from the history of science. Astronomers were checking Newton's theory against the calculations of the outermost known planet at the time, and the orbit of the planet was not where Newton's theory predicted it should be. According to Popper's system, finding one case falsified Newton's theory, and the theory should have been rejected. Rather than rejecting Newton's theory, however, the astronomers hypothesized that an undiscovered planet might be affecting the orbit of the outermost planet.

40

They calculated where such a planet would have to be, and they did indeed discover a new planet beyond the known planets.

This example raises a crucial question: when do we know that a result that differs from the predicted result is actually a case of the theory failing and not the result of an unknown factor? There is no way to answer this question—just as there is no way, according to Popper, to verify a theory. And so Popper's falsification criterion suffers from the same critique that he made against the verification system: in both cases the possibility exists that an unknown factor could change our conclusions. In the end, therefore, one false case does not necessarily falsify a theory.

Curiously, Popper anticipated this problem, but his answer was less than satisfying.[10] He argued that, in theory, his empirical method exposes to empirical testing *every conceivable way* a system could be tested and, therefore, provides a means of preventing an apparent false case like the example above. But how can we be so sure that we have included and tested all *unanticipated* factors? Popper's answer does not really address this issue.

THIRD CRITIQUE: *Common practice*

Popper's view seems antithetical to how we acquire and hold knowledge. We do not spend our time devising critical experiments to test all our cherished theories; we accept a theory unless some issue is raised that demands we revisit it. But even if we must revisit a cherished theory, the idea that we would naturally reject it based on one false empirical result—even from a crucial experiment—seems wrong. Popper values his empirical test too highly. It might be reasonable to shelve a result for a period of time, and negative empirical results do ultimately refute the truthfulness of a theory, but rejecting a theory is not always as easy or as mechanical as Popper claims.

To illustrate how Popper's view of testing hypotheses is too simplistic and mechanical, consider the case of D. C. Miller, which was brought to my attention in Michael Polanyi's book, *Personal Knowledge* (pp. 12-13). As is commonly known, Einstein based his theory of relativity in part on the results of the Michelson-Morley experiment. In a long series of experiments from 1902 to 1926, D. C. Miller and his associates repeated the Michelson-Morley speed-of-light experiments thousands of times with increasingly more sophisticated equipment, and yet each time they did the experiment their results contradicted the results predicted by Einstein's theory of relativity. Miller reported this in his presidential address to the American Physical Society in 1925. And yet, in spite of the fact that these experiments seem to falsify Einstein's theory, little attention was given to Miller's research. Physicists assumed that the results would be found to be in error at a later date. According to Karl Popper, this reaction was wrong; physicists should have rejected Einstein's theory immediately. But Popper has failed to understand the complex relationship between theories and fact, especially as that relationship is worked out in practice. To falsify a theory is not as easy or as objective as Popper thinks.

CONCLUSION TO THE CRITIQUE OF KARL POPPER

Just as we saw that the positivists' strong form of the argument for the objectivity of scientific knowledge is in error, we have now seen that the weaker form of the argument as made by Karl Popper is also in error. Although Popper (unlike the positivists) would acknowledge that values enter into the scientific process, he endeavors to preserve the unique status of scientific knowledge by insisting that the *testing* of hypotheses is value-free. Yet, not only would Popper's definition of science exclude many great scientists, neither logic nor practice support the idea that his method is uniquely objective.

SCIENCE IS NOT UNIQUELY OBJECTIVE

In the minds of many, empirical peer-reviewed science has priority over "subjective" areas, such as field experience, historical research, creative theory formation, and so on. According to this view, the privileged role of science arises from science's unique ability to separate fact from value by means of the scientific method. (This is assumption two of the three I listed at the beginning of the chapter.) As we have seen, however, neither the strong claim of the logical positivists nor the relatively weaker view of Popper holds up under scrutiny. No one can claim that science has a unique status because it deals with objective facts, because there are no such things as objective facts—that is, facts detached from any values. If the mind is active during observation, and observation is dependent on all our previously accepted beliefs, then there is no longer any foundation in which uniquely objective knowledge can take root. The scientist has no grounds to claim a unique certainty denied to other disciplines.

For subsequent discussion, I would like to highlight two important implications of this conclusion:

First, it is difficult to see how empirical science reported in empirical journal articles can have an exalted status as the ideal science. Once the notion that the scientific method yields objective knowledge is rejected, this empirical research has no inherently greater certainty, objectivity, or detachment than other types of science or any other endeavor to gather knowledge. The scientist's beliefs affect what he or she sees; and the beliefs, assumptions, and theories the scientist believes to be true determine the significance of the facts.

Therefore, scientific journal articles should have no higher standing than other forms of science (such as the discovery of a theory) or other forms of inquiry (such as history). (We will return to this issue in greater detail in the next two chapters.)

Second, if we believe that science is built one objective fact at a time, it may be difficult to determine which of two competing theories is true. Consider two astronomers observing a sunrise together. One holds the Ptolemaic theory, the other the Copernican theory. The Ptolemaic astronomer sees the sun rise, while the Copernican astronomer sees the earth turn. The sensations on their retinas are identical, but they see different things. No amount of looking at sunrises will resolve their differences. The empirical scientists and philosophers that believe observation is just opening one's eyes are wrong. Observation is not a passive mental process; there is a mental interaction between the sensations and the theory held by the scientist. Theories also in some sense establish the meaning of the data. (My purpose, however, is not to promote skepticism about our ability to know, as will become apparent in the next chapter.)

I do not believe, nor do I mean my critique to suggest, that empirical science is unimportant or indefensible. Empirical science plays an important role in science and, more specifically, in salmon recovery. Rather, my point is that the exalted status that radical empirical science has enjoyed is not defensible. My argument, then, is not against empirical science per se, but against a radical form that claims to be the ideal science and worthy of exalted status; neither one of these claims is true.

1. Edmonds, D. and Eidinow, J., *Wittgenstein's Poker* (New York: Harper Collins, 2001), 57.

2. Mach, Ernst. *The Science of Mechanics: The Apprehension of Facts* (London: Open Court Publishing, 1942) 5. Also quoted in: Hanson, N.R., Patterns of Discovery: An Inquiry into the Conceptual Foundations of Science (London: Cambridge University Press, 1969) 31.

3. Encyclopedia of Philosophy, (New York: Macmillan Publishing Co., Inc., reprint edition 1972 s.v. "David Hume."

4. Hanson, N. R., *Patterns of Discovery: An Inquiry into the Conceptual Foundations of Science* (London: Cambridge University Press, 1969), 11-15.

5. I do not want to take this critique too far. The critics are right that the concept of objective facts is in error, but in the next chapter I reject the skeptical conclusion that many of these philosophers of science saw implied by their critique.

6. I will return to this topic in chapter 6.

7. Kuhn, Thomas. *The Structure of Scientific Revolutions*, 2nd Ed. (Chicago: University of Chicago Press, 1970).

8. I agree with Thomas Kuhn that facts are not objective and as a result scientific knowledge cannot claim greater objectivity than any other endeavor. I disagree, however, with much of the rest of Kuhn's critique, but I will discuss the disagreement in the next chapter. In this chapter I have articulated Kuhn's criticism of logical positivism for two reasons: I agree with him that there are no such things as objective facts, and I want to provide background for my criticism of his view in the next chapter.

9. Jack Crabtree, a colleague at Gutenberg College, first presented this illustration to me.

10. Popper, Karl. *The Logic of Scientific Discovery*, 2nd Ed. (New York: Harper Torchbooks, 1968), 42.

CHAPTER IV: Were the Copernicans the founders of modern science?

"They use reasoning to overturn all reasoning,
and judge that they ought to have no judgment,
and see clearly that they are blind."

—Thomas Reid on skepticism[1]

The scientific community tells itself a marvelous story about the dawn of science, when the Copernican Revolution made it possible for mistaken and superstitious beliefs about the world to be replaced with real, objective knowledge about the world as it is. If, however, questions of science are as inevitably subjective as I have argued in the previous chapters, does that mean I believe we should be skeptical about our ability to know the truth of things? Am I trying to say that the inevitable subjectivity of knowledge makes any claim to knowledge suspect and nullifies any supposed accomplishments of the Copernican Revolution? Not at all. Part of my problem with the currently popular view of science is that it fails to adequately consider the problem of skepticism and, in fact, ignores the true lessons of the Copernican Revolution. Unwittingly, the emphasis on "objective," peer-reviewed empirical science has undercut the ability of science to achieve its own goals and has violated its own ideals.

My purpose in this chapter, therefore, is a little different from the previous two. In those chapters, my purpose was to refute two of the three assumptions that I find prevalent in the scientific community. And in the next chapter I will critique the third assumption, as well. At this point in my argument, however, I need to step back and get some perspective on the problem that I believe these assumptions have created for the modern scientific community. By embracing these assumptions, modern science is trying to maintain a position that is internally inconsistent and also inconsistent with the best part of its own heritage.

TWO QUESTIONS

Both philosophers and scientists are interested in these two questions: Can we know the truth of things? And if can know, how do we know? Let us consider three possible answers to these questions.

1. Skepticism

One possible answer is skepticism: it is not possible to know what is true about the world. As I mentioned above, my own position might initially be suspected to be a skeptical one; if all knowledge is inherently subjective, why should we believe any of it? Why I am not a skeptic will become clearer later; I will only say now that I deny that recognizing the inevitable subjectivity involved in knowing leads us to skepticism.

The philosophies that have led to our modern understanding of science, however, cannot equally claim to avoid skepticism as their outcome. The previous chapter discussed how logical positivism, Popper's hypothesis testing, and Kuhn's views on paradigm shifts were all ultimately skeptical:

1) Logical positivists based their system of radical empiricism on the epistemology of Hume, which was inherently skeptical.

2) Popper did not renounce the skepticism of the logical positivists. His method involved an empirical test based on the criterion of falsifiability: we can never claim a theory is true; we can only claim it has been severely tested.

3) Kuhn, of course, by the very nature of his argument rejected the idea of the objectivity of scientific knowledge. He saw no objective criteria by which to pick one paradigm over another; we can never say one paradigm is closer to the truth than another.

Thus we see that the dominant philosophies of science in the twentieth century all lead to epistemological skepticism. This fact is very striking. Yet, whenever I talk to scientists about how they justify their beliefs, they always respond (if they respond at all) with some version or combination of these three philosophies, and most commonly they respond with the ideas of the logical positivists or of Popper. You would think, therefore, that these scientists would follow their philosophical roots to their natural conclusion and be skeptical about our ability to know true things about the world. But that is not the case. Instead, almost every scientist I talk to would give a second possible answer to our two questions.

2. Empirically-based realism

My experience suggests that scientists are realists; they believe science tells them things about the real world. Yet, at the same time, they are strict empiricists. On the one hand, they believe in the superiority and objectivity of peer-reviewed empirical science—that is, they believe the three assumptions we have been discussing in this book, assumptions that grew out of the skeptical arguments of the logical positivists and Popper. But, on the other hand, they are not skeptics.

Frankly, I do not see how this can be. The only arguments I am aware of for the superiority of "objective," peer-reviewed science come straight from logical positivism and Popper. If those arguments are wrong (as I have argued) and if their skepticism is particularly unwarranted, then any justification for radical empiricism disappears. I can see no logically coherent way to be both a radical empiricist and a realist at the same time. Nonetheless, that is the position most scientists seem to have taken.

3. Empirically/subjectively-based realism

I would argue for a third answer to our two questions: a realism that recognizes the appropriate role of the subjective in knowledge. Empirical evidence is not bad; hypothesis testing is not bad; but in themselves they do not adequately account for how we know. We can come to know the real world, but it requires more than the strict application of the scientific method; it requires reason, field experience, theory construction, historical evidence, and all those other subjective elements that science often rejects as "anecdotal" or "gray literature."

I am concerned that by embracing the priority of peer-reviewed empirical science, the scientific community has at the same time unwittingly embraced implications that it does not believe. The rest of this chapter will flesh out what I mean.

THE COPERNICAN DEBATE

I will start at the inception of modern science. Examining the early development of science will help us understand the nature of science, its scope, and how its findings are justified. Because many of the assumptions of the early scientists differ from ours, examining them will help us both to illuminate our assumptions and to critique our position.

Because the Copernicans were instrumental in founding modern science, I will examine in some detail the debate that raged between the Copernicans and the Catholic Church concerning what science can know and what the grounds are for accepting that knowledge. In some sense, science has come full circle. I will argue that if today's empirical scientists and philosophers of science were to judge the facts of the Copernican debate, the overwhelming majority of them would side, surprisingly, with the Catholic Church against the Copernicans. This is particularly true of the empiricists and pragmatists who would most strongly reject the perspective for which this book argues. If I am right about the facts and their implications, the Copernican/Church controversy will help illuminate the central issues of my critique. In order for their view to be coherent the empiricists and pragmatists would need to re-write history and deny the Copernicans their place as the founders of modern science.

My reason for returning to this historical controversy, however, is not to lobby for a revisionist history. What interests me is that all three of the dominant twentieth-century philosophies of science (the logical positivists', Popper's, and Kuhn's) agree that science can tell us nothing about the world. If their formulation of science and its methods were used to judge the Copernican debate, the Copernican position would lose to that of the Catholic Church. While it is beyond my task to defend the Copernican approach to science (although I think I could), I believe the Copernicans were largely right and the Catholic Church was wrong. For my purposes, however, if I can just show that the logic of the modern position is, in fact, similar to that of the Catholic Church and thus against that of the Copernicans, then I will have shown there to be serious problems with the modern position.

I will address two specific questions: Were the Copernicans (in particular, Copernicus, Galileo, and Kepler) warranted in claiming that their theory was true? And if they were warranted, on what basis did they warrant their claim? Hindsight on that historical battle shows us that the Copernicans' claims were true. At issue, though, is how the Copernicans defined science and how anyone can claim to know something in science.

To help clarify the debate, I will describe (and refer to throughout the chapter) three likely positions that those involved in the debate held:

1) The Copernicans were not warranted in believing that their theory was true because the methods of astronomy and mathematics cannot prove what is true in the heavens. Their theory was only a method of calculating the appearances of the planets and the sun, a method that might be useful but one that the Copernicans could not claim to be true.

2) The Copernicans were not warranted in claiming that their theory was true because they had not demonstrated its truth. It is possible for the methods of astronomy and mathematics to prove the true motion of the heavens, but the Copernicans did not successfully demonstrate their theory to be true.

3) The Copernicans were warranted in claiming that their theory was true because they had provided sufficient evidence to support their claim.

Those advocating position one obviously believed there was no warrant at all for the Copernicans claiming they knew the true motion of the heavens. Those advocating positions two and three, however, agreed that proving the true motion of the heavens was possible, but they disagreed about whether the Copernicans had made their case. Why? Because they disagreed about what constituted a true warrant. As we will see in what follows, they emphasized (broadly speaking) either sense perception or reason as sufficient warrant for believing a theory to be true.

Now that I have described the three positions in the debate, let us look at the story of the Copernican Revolution as it is usually told today. I call it the "Copernican myth." Copernicus found astronomy on the brink of collapse, burdened by Ptolemy's complex geocentric system of epicycle upon epicycle. In a clean sweep, he replaced this geocentric nightmare with the simpler heliocentric system, thereby producing an accurate astronomy in far better agreement with reality. When Galileo took up Copernicus's cause, he ran into trouble with the obscurantist Catholic Church, which used the Scriptures and Aristotelian epistemology to try to defend its outmoded cosmology against the new scientific perspective based on sensory evidence and logical proofs.

Underlying this myth is the positivistic belief that science is the only form of knowledge. And so, according to the story, the Copernicans were warranted in claiming that their theory was true because, in keeping with the modern view of scientific knowledge, the Copernicans began with facts and then proceeded in a logical, inductive manner. In contrast, the Catholic Church grounded its opposition in religion, and there is insufficient warrant for religious belief (we are told) because faith is based on revelation and not ultimately on the evidence of the senses.

Upon examination, we will find that the Copernican myth is distorted in a number of crucial ways. None of the Copernicans were claiming that their heliocentric model made better predictions than those of the geocentric Ptolemaic system, nor were they claiming that they could prove their theory empirically. Nonetheless, they were claiming that their theory was true and that they had grounds for their claim. Furthermore, we will see that it is particularly ironic that so many twentieth-century scientists would champion the Copernicans, because none of the dominant twentieth-century philosophies of science (the logical positivists', Popper's, or Kuhn's) could agree that the Copernicans were justified in claiming their theory to be true.

So now, moving from the myth and keeping in mind the three positions on the debate that I described above, let us look at the Copernicans themselves and the views of the Catholic Church in order to answer our two questions: Were the Copernicans justified in claiming their theory was true? And on what basis did they warrant their claim?

COPERNICUS (1473-1543)

The first of the Copernicans, naturally enough, was Nikolaus Copernicus. Before we can talk about how he viewed the truth of and the warrant for his theory, we should clarify what he actually thought. Although it is true that he replaced Ptolemy's earth-centered model of the solar system with a sun-centered one, his thinking was somewhat different from what the Copernican myth portrays:[2]

1) The system of epicycles[3] in Copernicus's system was similar to Ptolemy's, and the system was quite rigid, not allowing for a number of additional epicycles.

2) Copernicus's system in no way increased the accuracy of predictions. Copernicus largely used Ptolemy's data.

3) The wanderings of the planets could be accurately accounted for by either system. Copernicus does not claim his system gives different empirical results.

We see therefore that Copernicus did not view his theory as simpler, better grounded empirically, or more accurate than the Ptolemaic system. This brings us then to our first important question: Did Copernicus believe his theory was true?

This is not an easy question to answer. The preface to his book, *On the Revolutions of Heavenly Spheres*, clearly indicates that Copernicus did not believe his theory was true and, in fact, ascribes position one to him: because an astronomer cannot by any means calculate the true movement of the planets, Copernicus was only proposing a new system of calculations for computing the apparent location of the planets.[4] But Copernicus did not write the preface. He only saw a printed copy of his work on the day he died. The preface appalled Galileo, Kepler, and several other followers of Copernicus. They clearly believed that Copernicus thought his theory was true—that he did indeed hold position three.

In his "Letter to the Grand Duchess Christina," Galileo claims that he should not like to have great men think that he endorsed Copernicus's position only as an astronomical hypothesis that is not really true.[5] And at the beginning of his *New Astronomy* (1609), Kepler, likewise outraged, identified the Lutheran theologian Andreas Osiander as the person who inserted the preface. The preface, therefore, is not strong evidence for answering our question.

In *Revolutions*, Copernicus claimed that he was led to search for a new system for deducing the motions of the sun, the moon, and the planets for no other reason than the fact that astronomers did not agree among themselves. After years of study he proposed the sun-centered solar system, and he concluded:

> Having thus assumed the motions which I ascribe to the earth later on in the volume, by long and intense study I finally found that if the motions of the other planets are correlated with the orbiting of the earth, and are computed for the revolution of each planet, not only do their phenomena follow therefrom but also the order and size of all the planets and spheres, and the heaven itself is so linked together that in no portion of it can anything be shifted without disturbing the remaining parts and the universe as a whole. (p. 508)

If Copernicus held position one as the preface indicated, then his position seems poorly reasoned. He would only have been claiming to have devised a new calculating method that yielded better results or was easier to use. Yet, nowhere does Copernicus make these claims, and so why should anyone prefer his calculating method to another's? Furthermore, the claim he does make about the coherence of his system does not seem relevant if Copernicus viewed his method as just a calculating device. The planets only *appearing* to move as a functional whole would have nothing to do with improved observations.

If Copernicus did not hold position one, did he perhaps

hold position two? Since his theory was no simpler or more accurate at prediction than the one it replaced, why would he recommend that other astronomers accept it? He sets out to put astronomy on a solid footing. But if he thought he had not proven his case, why would he even publish his book? It does not seem likely that Copernicus held position two.

We are left with position three, which indeed seems to be the position Copernicus held. The text of *Revolutions* leaves little doubt that Copernicus believed his theory was true and not just a computing devise. First, nowhere does he claim his work is merely a hypothesis. Second, throughout his work he describes the actual motions of the planets. Third, in his dedication to Pope Paul III, Copernicus claims he will demonstrate that the earth moves. Fourth, in the dedication, Copernicus ridicules the Church father Lactantius for speaking childishly about the earth not being a globe; and yet, if Copernicus had just been proposing another (among many) computing hypothesis, he would not have run the risk of offending the Pope by criticizing a Church father. And fifth, Copernicus's close friends, Bishop Tiedemann Giese and Georg Joachim Rheticus, were appalled by the preface to *Revolutions*.[6] There can be little doubt that Copernicus was claiming to describe the actual motions in the heavens.

Now we can turn to our second question. If Copernicus believed his theory was true, then on what basis did he believe himself warranted for thinking so? He did not base his claim on empirical observations of the individual planets, but rather he based it primarily on geometric reasoning (the arrangement of the orbit of the planets). He appealed to the coherence of the system, not to individual appearances of each planet. Copernicus's approach was an important shift in the methods of science. Although his system was empirically equivalent to Ptolemy's (nowhere did Copernicus claim his system yielded empirically different results or better predictions), Copernicus was claiming his theory explained the true motions of the heavens. He believed his theory was true, and the warrant for his claim was intellectual: his theory was *inherently rational*; "in no portion of it [could] anything be shifted without disturbing the remaining parts and the universe as a whole."

So then, we know Copernicus held position three, but what position in the debate did the Catholic Church hold? The Church did not respond to Copernicus immediately. He was, after all, a Church canon who had helped in the monumental task of revising the calendar. In a book published not long after *Revolutions*, the Dominican friar Giovanni Tolosani wrote that Copernicus's theory was absurd because it was scientifically unfounded and unproven.[7] But for over sixty years, besides Tolosani's critique, the Church paid little attention to Copernicus's theory and leveled few criticisms. We must look to Tolosani then to represent the Church's position. According to Tolosani, Copernicus did not have a physical theory from which he could deduce the earth's motion and neither had he presented a cause for its motion. Furthermore, in keeping with Ptolemy and Thomas Aquinas who believed astronomy could only be probable knowledge, Tolosani thought Copernicus mistakenly believed astronomy and mathematics could provide a basis for cosmology and physics. Ptolemaic astronomy, based on Aristotle's philosophy, said we cannot know the true motions of the heavens; and the Thomists said astronomy and mathematics could not provide information about the inference from the effects to the cause. Therefore, if Tolosani's position represented the Catholic Church's position in the Copernican/Church debate, then we must say the Church did not believe Copernicus was warranted in believing his theory was true. At this point, then, we can say the Church rejected position three, but whether the Church held position one or two is not clear.

GALILEO (1564-1642)

Next we come to Galileo Galilei, who was the lightning rod in the controversy between the Catholic Church and the Copernicans. What position did he take, and what position did the Catholic Church defend? To determine their positions, we will focus primarily on two formal events, often referred to as the first and second trials. Surviving documents pertaining to both events can help us ascertain both Galileo's position and that of the Catholic Church.

The first trial occurred in 1616. The Holy Office had decided the Copernican theory was foolish and absurd philosophically and that it was partly heretical and partly erroneous theologically.[8] The Church, therefore, issued two (conflicting) warnings to Galileo, but it did not prosecute him. The first warning forbade Galileo from holding the Copernican theory as true, but he could defend, teach, and discuss it as possible or even probable. The second warning forbade Galileo from holding, defending, or teaching the Copernican theory; but, it is important to note, he could still discuss it.[9]

The first warning, communicated to Galileo by Jesuit Cardinal Robert Bellarmine, was rooted in the epistemology of Thomas Aquinas, which makes a clear distinction between "true" knowledge and "possible or probable" knowledge. The first warning indicates, then, that the Church had placed Copernicus's theory into the category of possible or probable knowledge. In a famous letter to the Carmelite provincial Paolo Foscarini, Bellarmine said that if the Copernicans ever came up with sufficient proof that their theory was true, the Church would have to reinterpret Scripture; however, the Copernicans had not provided the necessary proof.[10] This is position two. Therefore, the Church allowed Galileo to defend, to teach, and to discuss the theory, but he could not hold it as true.

The second warning, coming from a faction within the Catholic Church, was rooted in a more skeptical epistemology. This faction, who strongly inclined toward mysticism, did not believe in the capacity of human intellect to attain truth. They held position one: they did not believe the Copernicans could ever know if their theory was true. Therefore they not only recommended that Galileo not hold Copernicus's theory as true, but also that he not teach or defend it (although he could discuss it).

The second trial occurred in 1633, soon after Galileo had published *Dialogue Concerning the Two Chief World Systems*.[11] The Church commenced proceedings against Galileo for violating the terms of the 1616 warnings. Again, within the Church there were two positions in the debate:[12] the Dominicans held position one, that the Copernican theory was to be understood as an abstract construct incapable of scientific proof; and the Jesuits held position two, that the Copernican theory was possible knowledge but that it had not been proven.

The Church leveled two charges against Galileo: first, that he had intentionally concealed the more severe of the two warnings from his publisher; and second, that in his book he had blurred the distinction between "true" and "possible" knowledge. The first charge is not germane to my argument. The second charge is relevant, however, because in answering it, Galileo identified himself with position two. He maintained that he had never held the Copernican theory as verifiably "true," but saw it as "probable" knowledge. He claimed that in an effort to be witty and clever, he had inadvertently made it appear that he accepted the Copernican theory as true. Galileo most likely believed the theory was true, but he apparently did not believe he had the necessary warrant to claim it was true. He penned this note in the preliminary leaves of his copy of the *Dialogue*:

> Take note, theologians, that in your desire to make matters of faith out of propositions relating to the fixity of sun and earth you run the risk of eventually having to condemn as heretics those who would declare the earth to stand still and the sun to change position—eventually, I say, at such time as it might be physically or logically proved that the earth moves and the sun stands still. (p. v)

Thus, while Galileo may have believed the Copernican theory was true, it appears he did not believe the Copernicans had yet proved it. Galileo clearly held position two. But what warrant would Galileo have accepted for justifying the truth of a theory? Would it have been grounded in sense perception or in intellect?

According to the modern Copernican myth, a warrant acceptable to Galileo would have been grounded in sense perception, that is, in observation of the heavens. After all, wasn't Galileo defending the new empirical science against the dogmatic Catholic Church? And wasn't the new science based on experience and observation, in contrast to the Church's outmoded epistemology grounded in Aristotle and the authority of Scripture? While there is some truth in the conventional story, there are some problems with it, as well.

First of all, the Church did not deny observation was important. Its Aristotelian-Thomist epistemology began with sense perception and then followed with abstraction and ultimately to a discussion of causes. How did this differ substantially from Galileo's methods? Furthermore, although the Copernican myth claims the Churchmen refused to look through the telescope because they were afraid of the facts— and perhaps this was true of some individuals—this was certainly not true of all. Some individuals merely considered the telescope to be a faulty instrument; they viewed it as unreliable because when it was set to different lengths, the appearances changed. Cardinal Robert Bellarmine accepted the telescope and looked through it with Galileo, but Bellarmine came to different conclusions. And a number of individuals were likely astute enough to understand that little or nothing was to be gained by the telescope observations themselves; because while telescopes could provide some supporting evidence for the Copernican theory, they did not show parallax in the stars caused by the movement of the earth. If the earth moved, observation of the heavens should show some evidence of it, but no movement was observed. This lack of parallax was a major stumbling block for the Copernicans at the time, and the use of the telescope only compounded the problem. But not all Churchmen were afraid to look through it.

Second, it is not clear that Galileo himself began with observation. Certainly, Galileo incorporated into the theory his observations obtained by using the telescope, but these observations were not conclusive; Galileo did not rely on them alone. For instance, in an important exchange of letters between Galileo and Kepler, Galileo wrote to Kepler that he had been a follower of Copernicus for a number of years and had indubitable proofs that the theory was true, although he had hitherto not dared to defend the new system in public.[13] When Kepler wrote back that he wanted to publish these proofs in German, however, Galileo broke off all correspondence with Kepler for twelve years. Furthermore, what observation did Galileo make that conclusively tipped the scales toward the Copernican theory? He did not point to a set of observations or to any one. So then, what role observation played in Galileo's science is not as clear as the Copernican myth sometimes assumes. In the *Dialogue*, the only positive argument Galileo tries to make for the Copernican theory is based on the tides (which was a wrong line of reasoning), and the rest of the *Dialogue* critiques the Aristotelian system.

It is hard to say what Galileo considered to be sufficient justification for the truth of a theory. It is not so clear what he meant by the "new way of observation and experiment." How do we reconcile the new method with the following famous quote from the *Dialogue*?

> Nor can I ever sufficiently admire the outstanding acumen of those who have taken hold of this [Pythagorean-Copernican] opinion and accepted it as true; they have through sheer force of intellect done such a violence to their own senses as to prefer what reason told them over that which sensible experience plainly showed to the contrary. For the arguments against the whirling of the earth which we have already examined are very plausible, as we have seen; and the fact that the Ptolemaics and Aristotelians and all their disciples took them to be conclusive is indeed a strong argument of their effectiveness. But the experience which overtly contradict the annual movement are indeed so much greater in their apparent force that, I repeat, there is no limit to my astonishment when I reflect that Aristarchus and Copernicus were able to make reason so conquer sense that, in defiance of the latter, the former became the mistress of their belief. (p. 328)

Also in the *Dialogue*, time and time again, it is Simplicio (the Aristotelian) who defends the role of the senses in the acquisition of knowledge and who complains that in the Copernican system the senses must be denied, and it is Salviati (Galileo) who wishes to deny not only the principles of science, but sense experience and the very senses themselves.

Galileo is famous for the "experiments" that he "performed," such as rolling round objects down frictionless planes or dropping differently weighted objects in the absence of air friction. However, it is difficult to say exactly what role these experiments played in Galileo's view of what warranted belief. These experiments are hardly what come to mind when an empiricist talks about "the facts" obtained from experiments. For Galileo, observation was not merely the quantification of bodies, but something else. And his experiments seemed aimed not at convincing himself of the truth of his beliefs, but at convincing others.

Galileo also continued the development of a mathematical description of nature, but he did not use Kepler's discovery of the elliptical orbits. Galileo may not have understood the significance of Kepler's work. Or, as some evidence near the end of the *Dialogue* suggests (Galileo complains that Kepler got caught up in the Aristotelian concept of the occult; p. 462), Galileo may not have accepted Kepler's view of science.

So, our question remains: What warrant would Galileo have accepted for believing a theory to be true? At this point, we can say only two things: first, Galileo's view of a proper warrant is complex and difficult to determine; and second, his view was certainly not the usual view of today's empirical science.

KEPLER (1571-1630)

Finally we come to our last Copernican example, Johannes Kepler, who published his *New Astronomy* in 1609. For a number of years, Kepler worked as an assistant to Tycho Brahe. Brahe had greatly improved the accuracy of the observations on the orbits of the planets, which Kepler used to discover these three laws:

1) The orbits of planets are elliptical in shape, with the sun at one of the foci of the ellipse.

2) The areas of sweeps of a planet within their planes are always equal.

3) The ratio between the square of the periodic time and the cube of the distance from the sun is the same for each planet.

There can be no doubt that Kepler held position three: he believed the Copernican theory to be true. And there can be little doubt that Kepler's warrant for believing the theory true was based on the inherent rationality of the three laws and their correspondence with the observational data of Tycho Brahe. Kepler, like Copernicus, believed human reason could come to knowledge of the heavens. Kepler also agreed with Copernicus that geometrical/mathematical reasoning was an acceptable method of science. This is not to say, however, that observation was unimportant to Kepler. Without the observations Tycho Brahe made, Kepler could not have derived his three mathematical laws; but his warrant for accepting the Copernican theory was always based on the inherent rationality of the mathematical models. History seems to agree with Kepler about the relative importance of theory over empirical data: after all, it is Kepler whom we recognize as the great scientist and not Brahe, who collected the data.

SUMMARY OF THE POSITIONS

Now that we have looked at three prominent Copernicans, we can summarize the positions in the Copernican/Church debate. With these positions in mind, we can look specifically at how the Copernicans viewed science.

The Catholic Church

The Catholic Church's position in the debate was divided. The Dominicans held position one: they believed that knowledge of the heavens was impossible. The Jesuits held position two: they believed that it was possible to know the true motions of the heavens but that the Copernicans had not proved their theory.

The Copernicans

Copernicus and Kepler held position three: they clearly

believed that the theory was true and that they were warranted in claiming it was true because the system was inherently rational.

Galileo likely held position three, but assigning him to only one position is difficult. He publicly claimed he held position two. But there is significant evidence to show he believed that the Copernican theory was true and that the Copernicans were warranted in their claim; if he did, then he held position three. We can say with some certainty, though, that Galileo did not believe strict empiricism to be the warrant in science.

The Copernican view of science

Copernicus and Kepler saw science as a search for the true theories that explain the actual motions of the heavens. They were realists, motivated simply by the desire to know the truth. They believed that an individual could use his or her reason to come to know the truth and that their theory corresponded to the true motions of the heavens.

This view of science was the true watershed in the rise of modern science. Before the Copernicans, science looked to Ptolemy and Aristotle, who believed we could never know the true motions of the heavens. (They would have held position one in the Copernican/Church debate.) The Ptolemaic-Aristotelian view limits science to a small role in the acquisition of knowledge of the heavens.

Copernicus and Kepler believed they could use reason to come to know the true motions of the heavens, and the warrant for their belief was probably their view of God. They believed God was rational and that he had created the universe to be rationally ordered and human beings to be rational.[14] And so the Copernicans believed they could use their rationality to know the truth about the created order—in particular, the true motions of the heavens.

Believing in man's rationality, the Copernicans did not

limit science to strict empirical methods and justifications. Others at the time also believed that knowledge of the true motions of the heavens was possible, but they defined science as limited to more empirical or Thomist-type proofs. Cardinal Robert Bellarmine, for example, took this position; he believed that the Copernican position was possible but that there was not sufficient proof for believing it to be true. In contrast, Copernicus believed his theory to be true based on the coherence between it and everything else he knew— in particular, mathematics. His theory was grounded in a geometric argument—not an empirical or Thomist argument based on the observable data. Both Copernicus and Kepler were relying on their rationality to arrive at a true theory, which Galileo recognized when he says (as I quoted earlier) that he cannot help but admire those individuals who hold the Copernican theory in spite of the data of their senses.

If science is limited to facts that can be empirically determined, then the Copernicans were wrong to claim their theory was true. This was the conclusion of Francis Bacon, the founder of modern empirical science.[15] He argued that the Copernicans were wrong to claim their theory was true, because they had no empirical evidence.

THE MODERN VIEW OF SCIENCE

Attentive readers may have noticed that the three positions taken during the Copernican/Church debate correspond roughly with the three epistemological options (skepticism, empirically-based realism, and empirically/subjectively-based realism) discussed at the beginning of this chapter. In the light of our discussion of the Copernican Revolution, let us examine those three options again.

Skepticism

Ironically, the modern day skeptics have returned to the position that some in the Catholic Church held. Proponents of all three of the dominant theories of science (the logical positivists', Popper's, and Kuhn's) would have agreed with position one in the Copernican/Church debate: the Copernicans were wrong to claim their theory was true; based on their experience at the time, they could not know the true motions of the heavens. And so modern day skeptics would join the ranks of Ptolemy, Osiander (who wrote the preface to *Revolutions*), and some Dominicans in the Catholic Church—all skeptics who believed we could not know the true motions of the heavens. The positivists would have held position one because they are radical empiricists who explicitly accept the epistemology of David Hume and who therefore believe we can have no "objective" knowledge of the world outside of our minds. Those subscribing to Popper's view of science as hypothesis testing would have held position one because, believing that we can only falsify hypotheses, they deny we can know that a theory is true. And Thomas Kuhn would have held position one; because of a problem he called "incommensurability," we cannot know if a particular theory (or paradigm) is closer to the truth than another theory. Anyone holding a conception of science that denies that science can come to the truth about the world would also have held position one—along with one faction of the Catholic Church and against the Copernicans.

The only philosophical justification the scientific community has ever been able to provide for its radical empiricism leads inevitably to skepticism. The scientific community needs to confront the implications of this. If scientists can never know the truth, then on what basis should science have any status as knowledge, let alone an exalted status? None that I can see. And why should anyone pay for scientific research if it is not going to lead to true theories?

If any of the skeptical views are right, then science is dead. According to Kuhn, science only moves from one paradigm to the next and we cannot know if we have made any scientific progress; and even while a paradigm is in place, the scientist's work is only puzzle-solving. We cannot know if we are closer to the truth. And the positivists and Popper give us no surer foundation for science since they maintain that we cannot know what is true. Why should anyone bother to do science? Science is not going to help us save salmon if it cannot lead us to true theories. Skepticism returns science to the cultural role it had during the Middle Ages.

Empirically-based realism

Most scientists would agree with what I have said about skepticism; skepticism is not an option for practicing scientists. They may be empiricists, but they are not skeptics. Yet, as I argued above, the position of the realistic empirical scientist is not intellectually coherent. I do not see how one can be both a radical empiricist and a realist; the philosophical justification for each precludes the other.

For the sake of argument, however, let us suppose that somehow a scientist could embrace the empiricism of the logical positivists and Popper while rejecting their skepticism. Ironically, this would bring the scientist to the place where he or she would have agreed with those in the Catholic Church who held position two—and so opposed the Copernicans. Anyone holding a conception of science that limits the warrant or justification of a scientific claim to strict empiricism would have denied the Copernicans' claim that their theory was true, because the Copernicans had not proven it empirically. Once sense perception is regarded as the sole basis for science, the Copernicans cannot be considered scientists. Again, this was Francis Bacon's claim against the Copernicans.

Modern science only investigates questions that can be addressed by strictly empirical methods. But because these methods aim to replace intellect and reason with sense perception and statistics, using them results in limited, unskilled endeavors. These limited, empirical investigations are then published as peer-reviewed journal articles. And most scientists think that accumulating these small steps will help them decide between competing theories.

The method of science promoted by the Copernicans was substantially different from the modern conception of the scientific method. Because the Copernicans were evaluating the truthfulness of two empirically identical theories, no amount of direct empirical research could resolve the controversy. No amount of looking at sunrises could resolve the conflict. Data could not decide the issue. The Copernicans' method relied on reason and geometry—in short, on intellectual reasoning.

The modern scientific community needs to come to terms with the implication of strict empiricism. By truncating science to a supposedly objective, peer-reviewed, empirical process, modern science has nullified the accomplishment of the Copernicans. If radical empiricism is the essence of science, then modern science needs to rewrite its own history because those in the Catholic Church like Bellarmine were right and the Copernicans made a serious mistake. Now, I cannot believe the scientific community actually believes that the Copernicans were wrong. So I would like to see the scientific community revise its current view of science to include once again as scientists people like Copernicus, Kepler, and later theorists like Isaac Newton.

Empirically/subjectively based realism

As must be apparent by now, I side with the Copernicans. Science is an enterprise where empirical data must take their place alongside the more subjective elements of reason and experience. Empirical data are important, but not uniquely so; Tycho Brahe's careful observations played a great role in the Copernican Revolution, but not a sufficient role. What modern empirical scientists seem to forget is that all data are linked to theories and to our previously accepted beliefs; we cannot investigate facts separated from their contexts. (That was the argument of chapter three.) The development of theories is as important as the collection of facts. And in fact, we recognize most of the great scientists for their development of theories, not for their accumulation of data. If science ever forgets that the theorist is more important than the data collector, that Kepler's achievement was greater than Brahe's, it will truly have lost its way.

1. Reid, Thomas. *The Works of Thomas Reid.* (1863; preface, notes, and supplemenatry dissertations by Sir W. Hamilton, Bart., 2 vols, 6th edition, Edinburgh: MacLachlan and Stewart and London: Longman 1863) 448a.

2. Gingrich, Owen. "A Fresh Look at Copernicus." In *The Great Ideas of Today,* edited by Robert M. Hutchins and Mortimer J. Adler. (Chicago: University of Chicago Press, 1973) 154-178. Also: Gingrich, Owen. "The Copernican Revolution." In *Science and Religion: A Historical Introduction,* edited by Gary B. Ferngren. (Baltimore: John Hopkins University Press, 2002), 95-104.

3. An epicycle is a hypothesized devise to try to describe the motion of the planets. It is a circle located on a circle. It originated in Ptolemaic astronomy and is based on Aristotle's assumption that all motion in the heavens is circular.

4. Copernicus, Nikolaus. *On the Revolutions of Heavenly Spheres*, 1543. Hereafter, I will refer to this work as *Revolutions*. All quotations from *Revolutions* are from the Great Books of the Western World Vol 16, edited by Robert M. Hutchins, (Chicago: Encyclopaedia Britannica), 505. All quotations from or references to Revolutions are from this reprint; page numbers will be cited in the text.

5. Drake, Stillman, trans. *Discoveries and Opinions of Galileo* (Garden City, NY: Doubleday Anchor Books, 1957), 167.

6. Dreyer, J.L.E. *A History of Astronomy from Thales to Kepler.* (New York: Dover Publications, 1953) 320.

7. Feldhay, Rivka. *Galileo and the Church: Political Inquisition or Critical Dialogue?* (Cambridge: Cambridge University Press, 1995), 205-208.

8. Feldhay, 26-52.

9. Feldhay, 45-52.

10. Feldhay, 29-44.

11. Galilei, Galileo. *Dialogue Concerning the Two Chief World Systems.* (1632; reprint, Berkeley: University of California Press, 1967). Hereafter, I will refer to this work as Dialog. All quotations from or references to Dialog are from this reprint; page numbers will be cited in the text.

12. Feldhay, 53-69.

13. Petersen, O. "Galileo and the Council of Trent," Journal of History of Astronomy (1983), 14:1-29.

14. Both Copernicus and Kepler over and over again claim that through their reasoning they are achieving knowledge of the order in the universe as created by God.

15. Jones, Richard F. *Ancients and Moderns: A Study of the Rise of the Scientific Movement in Seventeenth-Century England.* (New York: Dover Publications, 1961), 51.

CHAPTER V: Peer review:
A return to the medieval model of science

"Hence, I consider it not very sound to judge a man's philosophical opinions by the number of his followers."

—Galileo, in response to a critique that the Copernican theory must be a poor theory because it had attracted few followers[1]

Now that we have some historical perspective on the problem in science today, we can return to the three assumptions that this books sets out to critique:

1) Science ought to have a privileged position in salmon recovery.

2) Science deserves its privileged status because the scientific method gives science a certainty and objectivity denied to other pursuits, allowing it uniquely to separate facts from values.

3) The objectivity of science is further guaranteed by the formal peer-review process, which ensures the quality of the data.

From legislators to industry people to environmentalists, almost everyone believes with scientists that relying on "sound science" is the only way to recover Pacific Northwest salmonids. And they define sound science as "peer-reviewed science" because they hold assumption three. In chapters two and three, I refuted assumptions one and two. Now I will turn to assumption three and show that it, too, is fallacious. The formal peer-review process does not ensure the quality of the data, and neither does it guarantee the objectivity of science.

In chapter three, we saw that according to the prevailing view of science, strictly following the scientific method is the only way to obtain objective scientific facts. And in the eyes of many, peer review ensures adherence to the scientific method and therefore guards against erroneous knowledge.

A number of other scientists review a scientist's work to make sure he or she has carefully followed the scientific method. These reviewers corroborate that the scientist collected the data in an unbiased fashion, that he or she used the scientific method of analysis (the proper statistical test), and that the results follow from the analysis. Only then is the work accepted for publication. This process supposedly justifies the claim that the only knowledge we should accept for the purposes of recovering salmon and conducting watershed restoration planning is scientific knowledge; we should consider all other forms of knowledge as "gray" literature, or mere opinion. While the review process is not infallible, it ensures a high standard of accountability. And while scientific results are always open to further testing, the peer-review process ensures that scientific knowledge has inherently greater certainty and objectivity than knowledge acquired by other means. So the argument goes.

A number of problems beset this argument, however. My critique will focus on two questions: What is the purpose of peer review? and Where does the authority within science rest?

THE PURPOSE OF PEER REVIEW

According to the prevailing view, the purpose of peer review is to ensure the quality of the data, the proper use of the scientific method, and the validity of the work's conclusions before it can become part of the scientific literature. But is this purpose of peer review defensible? Let us consider the three questions below.

On what basis do we decide what constitutes "data"?

The problem with believing that peer review ensures the quality of the data lies in how the prevailing view of science defines "data": data are "objective" facts. These fundamental

74

units of science are nothing more than unbiased measurements, the result of carefully collecting information by rigorously following the scientific method. "Doing science" is accumulating these data, and scientific knowledge's privileged standing is rooted in the data's being objective.

Those who define "data" this way and see science merely as "data collecting" agree with the basic tenets of logical positivism—although they might not acknowledge it. In chapter three, I critiqued the positivists' perspective and showed that scientific data are not inherently more objective and certain than information from other pursuits, such as history. Virtually all philosophers of science have rejected the belief in objective facts (data) for the last half-century.

We saw that no objective observations (data) are independent of theory or other beliefs. If data do not refer to "objective" facts but rather to "theory-laden" facts, how is a science paper reviewed by, say, five scientists any better than a history paper reviewed by five historians? On what grounds can anyone possibly defend the science paper as the only form of knowledge? No justification other than the one I critiqued in chapter three has ever been given. Scientific knowledge is inherently no more objective and certain than other careful methods of knowledge acquisition.

What role do the beliefs of the reviewer play?

If, as I have argued, data are theory-laden, then it is a fiction that reviewers concern themselves only with data, method, and conclusions. Reviewers cannot help but evaluate the theories and beliefs of an author in the light of their own theories and beliefs. This is one of the correct and insightful portions of Thomas Kuhn's analysis. According to him, one of the purposes of peer review is to ensure that a scientist does not deviate from the current paradigm.[2] Consider the following example.

In 1979, I submitted a paper to the *Canadian Journal of*

Fisheries and Aquatic Sciences, published by the Fisheries Research Board of Canada. In it I presented a new predictive stream-classification scheme for use in the lower peninsula of Michigan. Three scientists, all well known, reviewed my paper: a fisheries biologist, who had written a stream-classification scheme for Ontario streams in the 1930s; a stream ecologist; and a geomorphologist. The fisheries biologist's review recommended publishing my paper with no changes, saying it was "the most significant advancement in stream classification in the last forty-five years." The stream ecologist basically said that the work was not worth the paper it was written on; it did not merit publishing. The geomorphologist said that while my work was a strange way to look at stream systems and was not completely compelling, he recommended accepting it with revision. As a good empiricist, I was dumbfounded by the responses. How was it possible that the same paper could elicit three radically different reviews? Needless to say, because of the second review, the paper was never going to see the light of day in that journal. These reviews had little to do with my actual methods and facts. The reviews dealt with "bigger" issues. For instance, the second reviewer did not believe stream classification was a useful pursuit, and it was unlikely that he would ever have accepted a paper on stream classification—no matter what its merits.

Whether a paper is accepted or not depends to a considerable degree on the "luck-of-the-draw." Had the stream ecologist not reviewed my paper, the journal might have published it. But either way, the review process said little about whether or not the paper was true. I can point to a number of peer-reviewed papers that contain major flaws. In the case of my paper, peer review mostly concerned the work's significance, which the reviewers judged based on their own theories of science.

What about truth?

As strange as it may sound, the way scientists understand peer review has little to do with the truth of a theory or hypothesis. That is because the peer-review process is subject to the same tension we discussed in chapter three. Its so-called "objectivity" is rooted in the inherently skeptical philosophies of the logical positivists and Karl Popper. For example, most scientists believe that scientific observation or experimentation is uniquely objective because it is based on sense impressions—just as the logical positivists argued. To talk about truth in this context is incongruous. A scientist either gets the expected result or does not; the impression itself is a fact. As we saw in chapter three, the positivists grounded their view of science on the epistemology of David Hume, whom we defined as a skeptic. Many scientists also accept Karl Popper's hypothesis-testing philosophy of science, according to which we can never prove a theory or hypothesis true; we can only falsify it. If we accept either the positivists' or Popper's philosophy of science, then we cannot claim a theory to be true. We see, then, that if peer review were truly the objective process that many think it to be, we would be left with no philosophical option but skepticism about its conclusions.

Once we have rejected the belief that theories can be true, we could opt for Thomas Kuhn's explanation of the peer-review process. He argues that peer review has less to do with facts than it does a scientist's role in the community of workers. When science is the work of a community of people, as Kuhn says, then communicating findings becomes a central feature of science. Peers determine a worker's credentials, which then determine the worker's place in the community. Publishing not only puts a worker's findings within the paradigm's body of scientific literature, but it also ensures the worker's place in the community. The role of peer review in publishing ensures that the worker's findings are compatible with the current paradigm. The work cannot be too controversial because that would upset the peer reviewer's place in the community. Often, then, the scientists who win the most prestigious awards, have the best academic positions, and publish the most papers are those who generate the most data and the least controversy.[3] Thus according to Kuhn, the peer-review process has nothing to do with the truth of a work; it has everything to do with a worker's place in the community.

If Kuhn's perspective is right, and if his skepticism is well founded (which I do not believe), then the only logical conclusion we can reach is that no grounds exist for believing peer review tells us anything about the truth content of a scientific work. We certainly can't appeal to the positivists or to Popper. So we are left with this question: Why is peer-reviewed scientific knowledge the only form of knowledge we can use to recover salmon?

Summary: so what is the purpose of peer review?

The peer-review process ensures the quality of the data, the correct use of the scientific method, and the validity of the conclusions from the analysis—this is the usual answer scientists give for the purpose of the peer-review process. Because this answer is based on logical positivism's false premises, however, no one can defend it as the purpose of peer review. No defensible purpose of the peer-review process has been clearly articulated. But without understanding the purpose of peer review, it is difficult to understand why or how the process warrants labeling a scientist's work "scientific knowledge" by including it in the scientific literature or why so many people accept scientific literature as the only form of knowledge.

THE AUTHORITY OF SCIENCE

The moderator at the 1998 Oregon Plan meeting concluded that the authority of science rests with the community of scientists through the peer-review process. And the moderator's conclusion reflects the consensus of virtually everyone involved in the salmon recovery process: authority rests in the scientific community. The community ensures that the body of scientific knowledge contains only the knowledge it accepts as authoritative, and it does so through the peer-review process.

I do not agree that the authority of science rests with the community of scientists.

To begin my critique, I will turn once again to history. During the transition from the Middle Ages to the Renaissance and the Enlightenment, the source of final authority shifted from the community of peers (the experts) to the individual. Luther's role in the Reformation and Galileo's role in the rise of modern science are two examples. The locus of science's authority became a central issue, especially in the Copernican/Catholic Church debate, as the Middle Ages declined and modern science rose. Who wore the final robes of authority for determining the truth of a theory? According to the Catholic Church, authority rested with the consensus of the community of natural philosophers. According to Galileo, however, every individual had the right to examine the evidence and decide for himself. Acknowledging this right was a significant, important, and good step; and because I do not believe many people would seriously disagree with this assumption, I will not defend it here.

I will return instead to the twenty-first century and ask this key question: Who now wears the final robes of authority concerning the truth of science? Those who insist on the peer-review process have placed these robes on the shoulders of a community of peers (the experts)—and have thus embraced a model of authority aligned more with the medieval Catholic Church than with the Copernicans. On the issue of authority, science has taken a giant step backwards. The "community of science" has replaced the Church's "community of natural philosophers." One priesthood has replaced another.

By returning to an older model of authority, those holding the prevailing view of science have denied the progress made by the Copernicans, and Galileo in particular. Galileo's reply to one of his major opponents, Sarsi (a pseudonym for Jesuit Father Orazio Grassi), reveals his model of authority. When Sarsi argued against the Copernican theory, he implied that it must be wrong because it had attracted so few followers; even sixty years after Copernicus had published his work, the overwhelming majority of natural philosophers did not accept the Copernican theory. He cited Jerome Cardan and Bernardino Telesio as examples of feeble philosophers who also had no followers, deducing from their inability to attract supporters that their theories must be lacking.[4] And so, Sarsi implied, if the Copernican argument were good, a large number of their peers would have accepted their theory, but since few had accepted it, the theory must be a poor one Here is Galileo's reply to Sarsi's deduction:

> Perhaps Sarsi believes that all the host of good philosophers may be enclosed within four walls. I believe that they fly, and that they fly alone, like eagles, and not in flocks like starlings. It is true that because eagles are rare birds they are little seen and less heard, while birds that fly like starlings fill the sky with shrieks and cries, and wherever they settle befoul the earth beneath them. Yet if true philosophers are like eagles they are not [unique] like the phoenix. The crowd of fools who know nothing, Sarsi, is infinite. Those who know very little of philosophy are numerous. Few indeed are they who really know some part of it, and only One knows all.

To put aside hints and speak plainly, and dealing with science as a method of demonstration and reasoning capable of human pursuit, I hold that the more this partakes of perfection, the smaller the number of propositions it will promise to teach, and fewer yet will it conclusively prove. Consequently, the more perfect it is the less attractive it will be, and the fewer its followers. On the other hand, magnificent titles and many grandiose promises attract the natural curiosity of men and hold them forever involved in fallacies and chimeras, without ever offering them one single sample of that sharpness of true proof by which the taste may be awakened to know how insipid is the ordinary fare of philosophy. Such things will keep an infinite number of men occupied, and that man will indeed be fortunate who, led by some inner light, can turn from dark and confused labyrinths in which he might have gone perpetually winding with the crowd and becoming ever more tangled.

Hence, I consider it not very sound to judge a man's philosophical opinions by the number of his followers. Yet though I believe the number of disciples of the best philosophy may be quite small, I do not conclude conversely that those opinions and doctrines are necessarily perfect which have few followers, for I know well enough that some men hold opinions so erroneous as to be rejected by everyone else. But from which of those sources the two authors mentioned by Sarsi derive the scarcity of their followers I do not know, for I have not studied their works sufficiently to judge. (*Discoveries and Opinions of Galileo*, pp. 239-240)

Galileo is claiming that the individual has the final authority to judge the truthfulness of a work. His critic placed the final authority for accepting a theory within the community of workers (the experts); he believed that their ignoring the Copernican theory for sixty years was evidence of its being a poor theory. But Galileo rejected the notion that the number of workers supporting a theory indicated its truthfulness or worthiness.

Sarsi spoke for the Catholic Church's position, and his logic is clear: the experts were in a position to know. The issues were difficult, and individuals outside the community of experts were unable to critique this difficult philosophical work because they had not been trained to do so. Because Copernicus had been trained in mathematics and astronomy but not physics and logic, for instance, he did not understand that mathematics and astronomy could not arrive at true knowledge unless they were grounded in physics and logic.

The logic used to justify the peer-review process today is similar to Sarsi's. If peer reviewers accept a work, it must be a good study and it is accepted into the body of objective knowledge. If they do not accept the work, it must not be a good study and it is not accepted into the body of objective knowledge.

Therefore, the position of the Oregon Plan meeting moderator and virtually everyone else involved in salmon recovery has more in common with the Catholic Church's position than the Copernicans'. The prevailing view of science assigns the final authority for deciding whether a theory is true or not to the community of scientists, and the community executes this authority through the peer-review process: if a work is not peer reviewed, it is not objective knowledge. It seems ironic to me that in the twenty-first century, many would advocate a position for science that has more in common with the medieval Catholic Church than with the Copernicans—especially when these same people would agree that moving the authority from the peer community (the experts) to the individual helped create modern science. They have simply replaced the Catholic clergy with a new scientific clergy. This is a giant step backwards.

A DISCLAIMER

At this point, I need to clarify my position. I am not saying that peer-reviewed science is entirely fallacious and without any merit. (Although I did agree with this pessimistic conclusion for a number of years, I now believe it is wrong.) After all, in the last one hundred years, science, largely through peer-reviewed journals, has achieved much. The problem I see rests largely with the expressed philosophy of science and its conclusions—not with the content of science. Scientists at their best are using a great deal of energy and skill to seek greater understanding of the world around them (more about this in the next chapter). But when they try to explain the philosophy behind what they are doing, and when they proclaim that scientific knowledge has greater standing than other forms of knowledge, I must object. No good scientist ever slavishly follows the mechanical scientific method. As we have seen, even if the scientist could, the resulting knowledge would not be more certain and objective than other "gray" literature. Good scientific research published in journal articles is valuable information; its peer review, however, gives it no greater warrant as knowledge than other well-researched "gray" literature has. Furthermore, as we have also seen, scientists who strictly rely on the mechanical scientific method, generally have poor judgment skills. In fact, the current process for accepting research for journal publication penalizes scientists who make judgments; results must be based on the proper statistical test—not on the judgment of a researcher. Based on my argument in chapter two, I believe science that makes its way into peer-reviewed journals is successful in spite of the best efforts of scientists and philosophers of science to destroy it. The claim that mechanical statistical analysis is superior to human judgment, however, is just wrong—a subject I will address in the last chapter.

CONCLUSION

If the Copernicans started science off with a solid foundation, how did it end up where it is in twenty-first century—rooted in skepticism and with a model of authority similar to that of the medieval Catholic Church? While this question deserves a book of its own, I will conclude this chapter with some brief observations that address it.

The road to skepticism in modern philosophy started with Descartes. He opened the door to it when he accepted the skeptics' demand that all our beliefs must be beyond doubt. Although Descartes's philosophy is not itself skeptical, it led to the philosophy of David Hume (which chapter three discussed). Once we reject the belief that the purpose of science is to establish what is true—that is, beliefs and theories should correspond to reality—we are left with few options. If we do not accept that human reason can discover truth, then we are left either with logical positivism or with some view of science similar to Thomas Kuhn's—a community of workers within a paradigm.

Logical positivism's false premises underlie the currently accepted purpose for peer review: to ensure the quality of the data (facts), the correct use of the scientific method, and the validity of the conclusions from the analysis. But we have seen that no one can defend this purpose; much more is involved in the formal peer-review process. This supposed purpose of peer review assumes that peer-reviewed papers contain the objective knowledge of science. But we have seen that this assumption is false, as well; all papers are grounded in broader theoretical and contextual issues.

Furthermore, peer review as it is practiced today places the authority of science into the hands of a scientist's peers. And we have seen that this model of authority rejects the Copernican model in favor of the medieval Catholic Church's model. Peer review as it is practiced today represents a giant step backwards in science.

At the beginning of this book, I set out to critique three assumptions underlying the prevailing view of science. We now see that none of the three can stand. Although they are

84

widely accepted in our culture, and specifically by those involved in salmon recovery, they are false. So now let us replace these false assumptions with true ones. In the next chapter, I will prescribe a way to save science and establish its true role in salmon recovery.

1. Galilei, Galileo. *The Assayer*, 1623. In *Discoveries and Opinions of Galileo*, translated by Stillman Drake. (Garden City, NY: Doubleday Anchor Books, 1957), 239-240. Galileo's more complete response is cited later in the chapter.

2. Kuhn, Thomas. *The Structure of Scientific Revolutions*, 2nd Editions (Chicago: University of Chicago Press, 1970), 164-70.

3. A similar point was made by Peter H. Duesberg in *Inventing the AIDS Virus*. (Washington, D.C.: Regnery Publishing, 1996), 67.

4. Sarsi, the pen name of Father Grassi (a Jesuit), has claimed Galileo is like two feeble authors who have no followers. These authors were Jerome Cardan (1501-1576) and Bernardino Telesio (1500-1588). Cardan was a noted mathematician and the author of works on philosophy, medicine, astronomy, and nearly every other branch of Knowledge. I do not know how his work was viewed at the time. Telesio advocated replacing the philosophy of Aristotle with a philosophy grounded only in sense experience. Ironically this is basically the modern view later advocated by David Hume and the positivists.

CHAPTER VI: Saving Science

The first five chapters critiqued the three major assumptions about science that scientists working on salmon recovery in the Pacific Northwest commonly hold:

1) Science ought to have a privileged position in salmon recovery.

2) Science deserves its privileged status because the scientific method gives science a certainty and objectivity denied to other pursuits, allowing it uniquely to separate facts from values.

3) The objectivity of science is further guaranteed by the formal peer-review process, which ensures the quality of the data.

None of these three assumptions is defensible. They reflect a belief deep within the modern psyche that objective mechanical standards result in greater certainty than human judgment. This is just not true. Clearing away the errors associated with "objective" science and its role in salmon recovery is the first step in saving both salmon and science. Now we must put science back on a firm foundation of true assumptions. In this chapter, I will propose four such foundational assumptions, I will explore their implications for salmon restoration, and finally, I will make a few recommendations.

A NEW SET OF ASSUMPTIONS

Almost everyone agrees that the Copernican Revolution was the first major milestone in the development of modern science. Although the conclusions of individual Copernicans were not always right, the Copernicans at least shared a common vision of science—a much better vision than the one built on the assumptions I have been critiquing in this book.

My purpose in proposing the four assumptions below is to set science back on the foundation the Copernicans constructed. I believe these assumptions can save science.

ASSUMPTION ONE:
We possess a considerable body of knowledge.

The first step on the path to saving science is to acknowledge that we already possess a considerable body of knowledge. We know a lot about the world (including salmon). If this knowledge does not exist (as the skeptics say), then science is reduced to merely a social activity characterized by adherence to a particular method; it serves no real end, and the scientific community has no purpose besides self-aggrandizement. The logic from the premise to the conclusion is inescapable.

The task of philosophy is to clarify what constitutes knowledge, what its limits are, and how we obtain it. Moderns and post-moderns, including pragmatics, will immediately object that I have just accepted the very thing that must be proved. But I do not think philosophy's task is to prove we have knowledge of the world. No sane person doubts having this knowledge. If a person truly doubts that he or she exists, that other people exist, and that an external world exists, then no philosophical discourse is going to convince that person—and he or she is best left to mental health workers, not philosophers. Although it is logically possible that we do not know anything about the world, no sane person lives life as a skeptic.

We must simply recognize that it is not the role of philosophy to answer the skeptics, because their question is not genuine. Skeptics do not for a moment live their lives doubting they have knowledge; they are merely playing mental games. When Descartes tried to answer the skeptics, when he accepted their challenge to doubt any belief that is not certain (that is, not logically necessary), he started philoso-
88

phy down the inevitable road to skepticism. I believe we must begin our philosophical investigation of knowledge by assuming we already possess knowledge about the world. All philosophical endeavors must assume something or they will never get anywhere.

We saw what happened in the Middle Ages when science was grounded in the skeptical epistemology of Aristotle and Ptolemy, who said that the true motions of the heavens were not knowable. Medieval science could give us a predictive model, but it could not speak to the actual motions of the heavens; it could not discover what it believed unknowable. Understandably, science received little attention during the medieval period; it was worthy of little effort because it could produce little.

Modern astronomy would not be possible without the epistemological shift in the foundation of science that Copernicus, Galileo, and Kepler demonstrated: they believed it was possible to know the true motions of the heavens. These men were realists, and they believed that science was rooted in the pursuit of truth. They all accepted the "correspondence theory" of truth—that is, they believed their theory corresponded to the true motions of the heavens. The basis of their belief was theism. Because they believed that a rational God had created both them and the heavens, the Copernicans also believed that their rational theory described the actual motions of the heavenly bodies. And they (Copernicus and Kepler, at least) believed they could use rational tools, like geometry, to discern these motions.

The critical shift the Copernicans demonstrated was away from skepticism to a belief that we have knowledge about the world. Without their belief that their theory corresponded to the way the objects in the heavens actually move, no science of the actual motions could have existed at all. I am proposing this same shift for us in the twenty-first century. We must move away from skepticism and embrace the assumption that we possess a considerable body of knowledge about the world.

ASSUMPTION TWO:
All knowledge is personal knowledge.

There are no such things as "objective" knowledge and "objective" facts. Rooting these two erroneous concepts out of our theory of science has been one of the major goals of this work. We have mistakenly believed that the heart of the scientific endeavor is to replace skill and art with objectivity and certainty. But (as we saw in chapter two) acquiring all knowledge is a skill—an art—that can never be reduced to a recipe. Therefore, all attempts to create "objective" science are fallacious.

New knowledge often begins with observation. But (as we saw in chapter three) observation is not the passive mechanical process David Hume and others articulated. Rather, observation is an active process, a skill that we must and do learn. When observing an object, we use skill and prior knowledge to move from sensations to knowledge of the object. This process is tacit and not conscious, but it is a true skill nonetheless.[1] Therefore, observation is an art that cannot be reduced to a mechanical process.

The sensations our senses receive "speak" to us like a language; and, as we do when we listen to someone speak, we move so rapidly from the sensations (the words) to what they signify (the speaker's meaning) that we are seldom aware of receiving the sensations at all; we just see what the sensations signify.[2] Consider the following example. Jane is walking down the beach during a storm. Ahead she thinks she can make out a rider on horseback, but as she gets closer she realizes that what she sees is just a piece of driftwood. Her mind received sensations and then worked actively on those clues in the light of everything else she knew. The process ended when Jane concluded that the object she saw was a piece of driftwood.

Since observation is a skill, the process can succeed or fail; Jane failed at first to "see" the driftwood. But if we are skillful, we must also be able to recognize mistakes and correct them. How did Jane recognize that she might be wrong about the rider on horseback? It would be difficult for her to articulate the process because she was not focusing on the sensations; rather, she moved immediately and tacitly to the object the sensations signified. Somehow she must have recognized that the sensations did not "fit"; they were not exactly those she would have expected if what they signified were a rider on horseback—that is, some of the sensations "resisted" being interpreted as a rider on horseback. And the closer Jane got, the more evident it became that, indeed, the sensations were different from those that would suggest a rider on horseback. This resistance and subsequent shift in what we see is largely outside of our conscious control. At a certain point we are satisfied that we are correct about what we see, but this point varies depending on how important the observation is or if it is unusual.

Jane's observation process is similar to your reading the words on this page. You move from the sensations (the printed words) to the meaning I am trying to communicate (what the words signify). In this case, the words are cultural conventions we have agreed upon (a facet of reading that is different from receiving sensations from the natural world), and we call the sum of our conventions "language." As you read these words on the page, you use your reading skill to move from the words to the meaning I am trying to communicate. Even though each word has a range of possible meanings, you have learned to sift rapidly through the possibilities and ascertain my meaning, just as you would immediately understand the two uses of the word "bank" if I said, "The First National Bank is on the bank of the river."

But how you can be certain that you are understanding

my (or any author's) meaning? This is a difficult but critical question. You believe you are understanding what an author is saying as long as you are "tracking" with the flow of ideas; that is, as long the meaning continues to fit with everything else you know, including the range of meaning of the words and what you know about the author. If you perceive some resistance, if some idea does not follow or seems to jump suddenly to a different thought-track, you become concerned that you may not be following the author's meaning. In the end, whether you in fact ascertain any author's intent depends on whether you understand the relevant background of the author's beliefs and whether all the words fit tightly into the author's meaning.[3] Thus, reading is a process similar to that of observation. Both are based on skills, and both rely on everything the reader or observer knows.

But not all new knowledge originates as observation. Sometimes it is a radical reorganization of the facts into a new picture or theory about some aspect of our knowledge. A new picture of the world, a gestalt-like shift, occurs. Often this process is not initiated by any new facts; rather, our minds reorganize known facts into a new comprehensive picture. The outcome is new knowledge, but it did not result from new information. Rather, this new knowledge resulted from mental activity, from skillful intellectual endeavor. Copernicus's and Kepler's achievements were of this sort.

Thus all our knowledge of the world is personal knowledge; we acquire it by using skills that we must develop and hone. Acquiring knowledge is not a mechanical process, nor can it be reduced to one. It is high art. All our efforts to reduce knowledge acquisition to a mechanical formula, therefore, are flawed.

ASSUMPTION THREE: *Gaining scientific knowledge, like gaining any knowledge, is an art.*

I said that all knowledge is personal knowledge, and scientific knowledge is no exception. While aspects of scientific knowledge differ by degree from other endeavors, scientific knowledge is not a categorically different kind of knowledge. All knowledge is personal knowledge, acquired through art and skill; therefore, gaining scientific knowledge must also be an art and grounded in the skills of its practitioners.

ASSUMPTION FOUR: *What characterizes scientific knowledge is not its method but its subject matter.*

Most scientists today, including the moderator of the Oregon-Plan meeting who inspired this book, are "objectivists." They believe that the scientific method (described in chapter two) makes scientific knowledge unique because the knowledge that results from its use is more certain and objective than any so-called knowledge that might result from other endeavors. Thus they define science by its method: using the scientific method is "science," and only knowledge gained by using the scientific method (as reported in peer-reviewed journals) is "real" knowledge.

With assumption four, I am proposing that we no longer define science by one method of investigation. Rather, we should define science by what we are investigating. Science is the investigation of the physical world, with the goal of establishing what is true about the physical world so that our beliefs and theories correspond to reality. And just as we can attack any problem in a number of ways, so scientists can use a wide range of methods (both intellectual and experimental) to investigate the physical world; no one way always results in more certain and objective knowledge. I am proposing assumption four for three reasons:

1. The role of theory

The narrow definition of science in terms of one "scientific method" has lost sight of the central role that theory plays in the acquisition of knowledge. Science is always a mix of theory and experimental research. On the one hand, experimental data is always evaluated in the light of our current theories. On the other hand, our current theories can undergo radical shifts based on experimental observations. This is a normal part of the way we come to know things.

Unfortunately, modern science, rooted as it is in both logical positivism and Popper's hypothesis testing, has downplayed the inevitable interaction of theory and data. As we saw in chapter three, the positivists tried to reduce theory to a collection of facts, and Popper tried to reduce theory to being vulnerable to one falsifying fact. Neither of these approaches is defensible.

By defining science in terms of its subject matter rather than its method, we leave room for theory to play the essential role that it must play. In the interplay between theory and data, theory is the more important partner. The individuals that we recognize as great scientists were more theoreticians than hypothesis testers. The Copernican Revolution did not result from new data or more accurate predictions; rather, it was an intellectual reformulation of the existing theory of the heavens. The modern objectivist view, therefore, is at its heart anti-intellectual. By emphasizing the method, it has forgotten where some of its greatest strengths lie. No definition of science as a method can begin to capture the range of individuals and experiments that are scientific.

2. The importance of skill and experience

The narrow definition of science in terms of one "scientific method" has lost sight of the role of skill and experience in diagnosing problems and prescribing solutions. As I argued in chapter two, it is the general practitioner and not the research scientist that we go to for medical advice. Science should be defined in a way that leaves room for the vital role of skill and experience.

3. The nature of truly relevant data

The narrow definition of science in terms of one "scientific method" has lost sight of what ought to constitute relevant data. By emphasizing the supposed objectivity of the scientific method, we have lost sight of the variety of kinds of information—derived from experiments, from history, from logic—that we need to understand the world.

In assumption four, then, I am calling for a new view of what constitutes science because the old view has lost sight of the role of theory, the importance of skill and experience, and the nature of truly relevant data. The definition of science that focuses on the scientific method has truncated the process of how we come to know things. In so doing, it has crippled science's ability to solve the problems it addresses. To follow the modern definition of science to its logical conclusion would be the death of science.

SUMMARY OF THE FOUR ASSUMPTIONS

These, then, are the four assumptions I believe can restore science to the foundation the Copernicans established:

1) We possess a considerable body of knowledge.

2) All knowledge is personal knowledge.

3) Gaining scientific knowledge, like gaining any knowledge, is an art.

4) What characterizes scientific knowledge is not its method but its subject matter.

Replacing the three commonly-held assumptions that I have shown throughout this book to be fallacious with these four new assumptions gives us a firm foundation for all scientific pursuits, including the problem with which we began: salmon recovery.

IMPLICATIONS FOR SALMON RECOVERY

A number of critical implications for salmon recovery result from this reconstruction of science—from the scope of information we need to consider, to the accepted methods of evaluating that information, to the credentials of those making judgments about salmon restoration and its successes or failures. To illustrate some of the implications, I will return to the four examples in chapter one that I used to show the results of applying a fallacious understanding of science to salmon recovery. And with the new assumptions in mind, I will suggest what right actions could have been taken in those situations—actions that would have better benefited salmon.

Examples one and two

The first two examples from chapter one illustrate the error of assuming that we can only use scientific information in the salmon recovery process. In the first example, the National Marine Fisheries Service used only the information on coastal cutthroat trout from the Winchester Dam site on the Umpqua River to establish whether the populations were depressed from historic levels. In the second example, the Pew Charitable Trust's "blue-ribbon" science panel could not establish with any certainty that the salmon populations were depressed from historic levels. In both examples, scientists were trying to use the methods of science to answer a historical question: What were the historical populations of salmon in Pacific Northwest rivers? But the methods of science are not applicable to this question; and, as these examples show, it was a categorical mistake to assume that only science could make the assessment.

This mistake is a holdover from logical positivism and the belief that science is the only form of knowledge because the scientific method supposedly separates facts from values.

But as our two examples show, we need historical information to answer questions about salmon runs in the past. We need experienced natural historians—not empirical scientists—to examine historical evidence and to make the necessary judgments. Here, empirical science should play a subservient role (at best) in the process.

And if it is a mistake to assume that only science can make assessments—as it was in the two examples—then it is also a mistake to assume—as the current model of science does—that peer-reviewed journal articles should have a higher status than so-called "gray" literature. As I argued in chapter five, this view returns science to a medieval model. We need to consider a broad range of information, all potentially of equal value. While I would agree that not all information is equally valuable, its value does not depend on where it is published, whether it is science or history, or who reviewed it. When we ask questions about the past—what watersheds looked like and what the size of salmon runs were, for example—we need to consider historical information. Good natural history is the foundation for what we know about the natural world.

Examples three and four

The third and fourth examples from chapter one illustrate the error of assuming that only empirical scientists are experts on questions relating to science. In the third example, the Governor of Oregon appointed a Independent Multiple Disciplinary Science Team (IMST) to evaluate whether the Oregon Plan to recover salmon was working, thus implying that only scientists could determine the plan's effectiveness. The idea that a panel of scientists wears the ultimate "robes of authority" with regard to truth is a fallacious holdover from positivism. The team should at least include good natural historians and philosophers of science (among others) because a broader range of expertise is needed to evaluate the

Oregon Plan's effectiveness than empirical scientists alone can provide.

In the fourth example, the president of the Oregon Chapter of the American Fisheries Society, responding to a bill introduced into the Oregon legislature defining science as "hypothesis testing," recommended that a scientific panel establish what is good science. The idea that scientists ought to determine what is good science is another holdover from positivism. We need a panel of philosophers—not scientists (or politicians)—to address the question because the methods of science cannot help define sound science and its accepted methods.

In general, we need to ask who has the requisite experience to make the best judgments regarding the critical issues facing us in salmon recovery. Individuals placed on scientific or other panels should be chosen for their ability to render judgments, not for their published record. As I argued in chapter two, making judgments is a skill; but science as it is currently construed works against scientists developing this skill and instead replaces judgment with statistical tests.

The best possible scenario, then, for ensuring that the panels thoroughly review all information relevant to salmon recovery and make sound judgments based on this information is to make sure that a majority of the people on the panels have years of field experience dealing with salmon. Questions that require judgment about salmon ought to be answered by individuals who have direct experience with salmon. This is analogous to the medical example I cited in chapter two: the general practitioner who has years of experience making diagnoses (judgments) is the best person to render a diagnosis, not the medical researcher who replaces judgment with statistics.

We need to replace the belief that the community of scientists wears the ultimate robes of authority for determining what is true. I argued in chapter five that this belief returns science to the medieval model practiced by the Catholic Church, in essence making the community of scientists into a new priesthood. Scientists are not inherently better than any other group at making critical judgments; in fact, in many cases they are making judgments that they have no business rendering—yet another holdover from positivism. Galileo's comments to his detractor Sarsi ring loud and clear: the foundation of modern science is the right of the individual scientist or philosopher to stand for what his own reason tells him is true. Ultimately, this is science's only true ground.

RECOMMENDATIONS

Since making judgments about salmon recovery is a skill rooted in personal knowledge (as we saw in chapter two), we need criteria for evaluating the ability of those involved in salmon restoration to make those judgments. Some people have natural talent and are quick learners; others take longer to learn, but they gain proficiency as they gain experience. Before we put anyone in the position of making judgments about salmon recovery, however, we should determine what level of skill the person has.

But how are we to make such a determination? Because I am a pilot, I will use the Federal Aviation Administration (FAA) as an example. The FAA, which considers piloting an airplane a skill and not just a matter of reading a how-to manual and performing the required tasks, has instituted a number of criteria to ensure that pilots have skill and maintain it. Any skill takes constant practice and use; a person loses a skill he or she does not use. In order to carry passengers, therefore, I am required by the FAA to take off and land at least three times during each ninety-day period; and to stay current, I am required to make a certain number of instrument approaches each year. Recognizing the limitation of using these criteria alone, however, the FAA also requires me to take a biennial, instructor-evaluated flight test. The ultimate test of my piloting skills, then, is how well I can maneuver the plane, just as the best measure of any skill—ultimately, the only measure—is its outcome.

Therefore, to determine how skilled a person is at making judgments regarding salmon recovery, we need to examine his or her work in this area. We ought to be able to examine the scientist's field journals and completed restoration plans to see both how much time the scientist has spent in the field and whether he or she has constructed good salmon-recovery plans. Because the skills of watershed analysis and restoration planning take time to develop and to maintain, the basic measure of "field experience" should be the amount of time spent doing watershed analysis and restoration planning—not the time spent collecting information for one's own research. Time spent in the field with graduate students, time spent showing fieldwork to others ("dog-and-pony shows"), or time and effort provided by technicians should not count toward a scientist's field experience.

In addition, we ought to examine the outcome of the scientist's previous judgments relating to salmon recovery. Has the scientist recognized errors and learned from them? Has the scientist made good judgments—in so far as time has shown the effects of his or her decisions? Admittedly, answering this second question can be difficult because the effects of many judgments take decades (or even centuries) to manifest themselves. In the 1960s and 1970s, for example, fisheries biologists decided to spend most of their habitat-restoration money on removing logjams from streams—an activity which a decade or so later was realized to be one of the most destructive to salmon in the Pacific Northwest. The biologists' decision was not based on sound field experience and therefore had negative consequences. Although assessing the long-term effects of some judgments may be difficult, it

should not deter us from examining the results of a scientist's work to the extent possible.

To evaluate the skill of an individual is more difficult than to rely on an objective standard such as how many peer-reviewed journal articles a person has authored. Critics, therefore, will reply that the criteria I am suggesting are unworkable: the chaotic result would severely impede restoration efforts. The supposedly objective standard, however, is no guarantee of success. I have shown it to be based on fallacious reasoning.

CONCLUSION

I began this book after hearing the moderator at the 1998 Oregon-Plan meeting express a commonly held view of "sound" science. But although the view is commonly held, I have shown throughout this book that its foundational assumptions are false. My goal has been to show how these assumptions hurt science and how they hurt salmon recovery in particular. But I have also tried to point the way to a view of science built on true assumptions and to show how such a view can not only save salmon, but more importantly, how it can save science.

I fear, however, that we will continue to convene more scientific panels composed of individuals selected on the basis of their peer-reviewed work. Although the goal of these panels will be to establish what is sound science and to make recommendations for salmon recovery, many of the panelists will have little or no experience on which to base their judgments about these important issues. The future of science and salmon will suffer as a consequence.

Skill at making judgments should be the main criterion for placing individuals in positions to make critical decisions. We need general practitioners—not researchers—in the critical decision-making positions that require judgment.

Although scientists can meet this criterion, their view of science and its methods can seriously impede their judgment; indeed, the extent to which a scientist's judgment is impaired can be measured by the extent to which he or she believes that the mechanical scientific method gives privileged knowledge. If scientific knowledge is the only form of knowledge admissible, then no judgments whatsoever are necessary or desired, because judgments lie outside of science.

The Copernicans initiated modern science with a very different set of assumptions than those of twentieth-century empirical science. In order to save science, we need to return to a model of science much closer to the Copernican model. If we do not, I fear that the results will be bad for science and bad for salmon.

1. Polanyi, Michael. *Personal Knowledge* (Chicago: University of Chicago Press, 1957), 95-100.

2. This view of observation relies heavily on the works of Thomas Reid and Michael Polanyi.

3. Three of my colleges at Gutenberg College (Ron Julian, J. A. Crabtree, and David Crabtree) have published a book on this topic. While this book, *The Language of God: A Commonsense Approach to Understanding and Applying the Bible* (Colorado Springs: NavPress, 2001), applies specifically to biblical interpretation, it has much to say about how language works in general and how in any literature we come to know an author's intended meaning, which is the goal of reading.